I0043489

Charles A. Stephens

Off to the Geysers, or, the Young Yachters in Iceland

As recorded by "Wade"

Charles A. Stephens

Off to the Geysers, or, the Young Yachters in Iceland
As recorded by "Wade"

ISBN/EAN: 9783337317966

Printed in Europe, USA, Canada, Australia, Japan

Cover: Foto ©berggeist007 / pixelio.de

More available books at **www.hansebooks.com**

THE CAMPING-OUT SERIES.

VOLUME III.

OFF TO THE GEYSERS:

OR,

THE YOUNG YACHTERS IN ICELAND.

AS RECORDED BY "WADE."

By C. A. STEPHENS.

ILLUSTRATED.

THE JOHN C. WINSTON CO.,

PHILADELPHIA,

CHICAGO,   TORONTO.

Entered according to Act of Congress, in the year 1873,

By JAMES R. OSGOOD & CO.,

In the Office of the Librarian of Congress at Washington.

INTRODUCTION.

BAHIA, BRAZIL, Dec. 29, 187·

THE EDITOR OF "OUR YOUNG YACHTERS' SERIES."

Dear Sir, — We would earnestly wish to avoid giving the impression that our "method of self-education," as you have been pleased to term it, consists merely of a series of yacht-cruises off and on. This would be an error. Both myself and the young gentlemen associated with me recognize, as rigidly as the most rigid of our university-professors, the necessity of a thorough drill and discipline of the mental powers. In proof of this, allow me to acquaint you somewhat with our last season's studies. Fully a third of the time from November till April was occupied in making calculations connected with an actual survey which we made of Massachusetts and Cape-Cod Bays. We even went so far as to get "regular

lessons " from Greenleaf's " Intellectual Arith-
metic," and add up triple and quadruple col-
umns of figures according to the " accountant's
method : " this for mental drill.

I remember, too, that we used to multiply, *men-
tally*, long arrays of figures by 13, 17, and 19 ; as,
for example, $9783216354783 96 \times 17$.

During this time we read " Heat, a Mode of
Motion," and " Fragments of Science," by Tyndall ; " The Descent of Man," and " Origin of
Species," by Darwin ; Prof. Huxley's " Lay Ser-
mons ; " and also the first four volumes of Ban-
croft's " History of the United States." These,
besides *papers*, *magazines*, &c.

I call your attention to these studies, only to
illustrate to you our " mode of self-education."
In it the *yacht* figures but as a single feature. We
have become " yachters " to avoid the tedious,
vegetable life of a college.

We have thought your Introduction to the second volume * liable to mislead. Set us right before
your readers in this particular, if possible.

Very truly and respectfully yours,

J. W. RAEDWAY.

[We know of no better way to make the correction than by
printing this letter as an introduction to our third volume. —
Ed.]

* " Left on Labrador."

CONTENTS.

v

CHAPTER IV.

CHAPTER V.

CHAPTER VI.

CHAPTER VII.

CHAPTER VIII.

CHAPTER IX.

CHAPTER X.

CHAPTER XI.

LIST OF ILLUSTRATIONS.

OFF TO THE GEYSERS.

CHAPTER I.

'Off to the Geysers."—Wash's Letter.—The Sailing of "The Curlew."—
The Cold Storms of the North Atlantic.—The Atlantic Cable.—Over
the "Telegraphic Plateau."—The Bottom of the Ocean.—Honor to
Mr. Field.—An Iceberg.—A Whale.

"COME on to-morrow. We're off for the geysers."
This telegram came to hand the 17th of May,
the spring after our voyage to Hudson Straits, which
forms the subject of Wash's narrative of "Left on Labrador."

I was in Baltimore on a visit to my mother and sisters
at this time, and had not been present at the "council"
which the boys had held about a fortnight previous.
Wash had kindly kept me posted, however. The plan
for the summer cruise will appear from the following ex-
tract from one of his letters received a few days before.
Under date of May 3 he writes, —

"Held our final grand powwow yesterday. Gave up
the idea of trying to go into Hudson Bay again, — this

1

season, at any rate. Raed then proposed Greenland, — the southern and south-western coast from Juliana-shaab to Upernavik. His idea was to sail direct for Julianashaab, and, as the season advanced, make our way as far north as Upernavik; examining, as we went, the fiords, or bays, and the great glaciers that move down into them from the mountains, and break up to form icebergs.

"Greenland, he argued, is the home, the nursery, of icebergs. Let's examine it thoroughly. Ice has been in the past, and is still, to a great extent, the agent by means of which the rocks have been *ground up* to make soil. Let's go see for ourselves how this is done.

"But Kit thought we had better go to Iceland to see the geysers and the fire-jokuls, and study the geology and mineralogy of that strange volcanic island. There, too, were the quaint and curious *sagas*, — stories of the rough Northmen of early days. How impressive and pleasing to hear these tales from the mouths of Iceland-ers, who are said to spend the hours of their long winter nights poring over these books, never tiring of their wonders !

"Well, we had quite a discussion. Wish you had been there ! The *pros* and *cons* of both propositions were set forth at length. Come to talk it over, we all three came to like the Iceland plan best, and voted for it. So Ice-land it is, this summer. Hurrah for the geysers !

"Now hurry up, Wade. Get ' mamma ' visited as soon as possible. We are busy as beavers getting ready. We're having ' The Curlew ' fixed up nice, painted new; and we've got a bonny new silk flag, — a beauty ! You know we're in funds just now.* Capt. Mazard is seeing that every thing about the schooner is put in tip-top, apple-pie order. We can't get Donovan this year: he's gone up to the Banks, master of a cod-fisher. Pretty good for Don. Wish we had him, though! Remember how he slung those Huskies round for us? Old Trull, too, has gone back to the navy: seems to me it's 'The Franklin' he's on now. Poor Corliss, you recollect, was lost from a brig last winter. But we've got Weymouth, Hobbs, and Bonney left us, besides Palm-leaf and Guard. We have hired four new men. Two of them, Smith and Trevers, are from Gloucester. Another, by the name of Elder, is from Nantucket. Then there's a New-Bedford chap who calls himself Truax: funny name, that!

"We are getting in a better lot of 'grub' than we had before. Mean to live better. Then we have bought a nine-by-eleven tent: that is for land-travel after getting there. We are going to take all of our great-coats and things that we had last season, as well as about a thousand dollars in money, — gold. Come now; be on hand.

"My duty to your mother, and my warmest regards to my pretty cousins, whom I hope to have the pleasure of seeing next winter.

"Yours, "WASH."

This explains it as well as any words of my own would do. I set out for Boston the next morning after getting

---

* Our ten thousand dollars which had been invested in Back-bay land had been sold out a few weeks previous for seventeen thousand three nundred and sixty dollars

the despatch, and arrived at daybreak the morning after.
This was the 19th instant. We sailed on the afternoon of
the 20th. Had favorable weather, with the exception of
one heavy shower and squall the night of the 23d, till
the morning of the 25th, while off Cape Race, when we
encountered a strong north-easter for thirty-six hours.
A cold, stormy time. All our great-coats and blankets
were needed to keep us comfortable. Comfortable, did I
say? — to keep us from freezing, I had better said.
Withal, we were a little seasick. Oh these North-At-
lantic storms and chilling fogs, and blasts fresh from the
lair of icebergs! — they are enough to chill the very
marrow in one's bones. But with a stove as hot as we
could get it for the interrupted draught, buffalo-skins,
wool-blankets, and plenty of hot coffee, we managed to
live. These discomforts we expected. Our last year's
voyage had taught us how the North Atlantic looks and
feels.

A stanch craft, and sailors practised amid its sleet-
storms and inured to its colds, are only able to brave its
reckless black billows, that forever war against the ad-
venturous mariner.

On the morning of the 28th we were in latitude 48°
35′, longitude 25° 57′ east of Washington, or 51° 3′ west
of Greenwich, — about a hundred and fifty miles east
of Newfoundland.

"Right over the Atlantic Cable," Raed remarked.
He had the chart spread out on the table. "The New
World and the Old are talking with each other along
this little cord lying somewhere under us, down amid
the cold, black ooze of the ocean-bottom."

"What a grand thought that is!" exclaimed Kit,
— "two nations conversing easily and instantaneously
across this stormy black ocean! Two thousand miles
of wild, wrecking waves crossed in a second! It's the
greatest achievement of this century, or any century!"

"How deep under us do you suppose the cable lies?"
Wash queried.

"Not over a mile and a half here," Kit said. "What
they term the Telegraphic Plateau is at no point between
Newfoundland and Ireland much over twelve thousand
feet under water. That is why they call it a plateau, —
because it is a plain raised higher than the ocean-bed
either to the north or the south of it. To the south-
east of the Grand Bank, the Atlantic is nearly six miles
deep; and, not three hundred miles to the north of us,
the depth is scarcely less than four miles. But according to Lieut. Berryman of our navy, and Capt. Dayman
of the Royal Navy, the depth along the telegraphic
plateau nowhere exceeds twenty-four hundred fathoms, —
fourteen thousand four hundred feet."

"Here's Dayman's report," remarked Raed, opening
our snug little bookcase (for during the last six months
we had been getting together quite a library of scientific
books and reports); "and here's what he says: 'This
space has been named by Maury the Telegraphic Plateau;
and although, by multiplying the soundings upon it, we
have depths ranging from fourteen hundred and fifty to
twenty-four hundred fathoms, these are comparatively
small inequalities in its surface, and present no new
difficulty to the project of laying the cable across the
ocean. Their importance vanishes when the extent of

the space over which they are distributed (thirty degrees of longitude) is considered.' "

"Here's something else on the same subject," said Wash, taking down another book, — a "History of the Atlantic Telegraph," I think. "Hear this: 'The ocean-bed of the North Atlantic is a curious study; in some parts furrowed by currents, in others presenting banks, the accumulations, perhaps, of the *débris* of these ocean-rivers during countless ages. To the west, the Gulf Stream pours along in a bed from a mile to a mile and a half in depth. To the east of this, and south of the Great Banks, is a basin, eight or ten degrees square, where the bottom attains a greater depression than perhaps the highest peaks of the Andes or Himalaya : six miles of line have failed to reach the bottom. Taking a profile of the Atlantic basin in our own latitude, we find a far greater depression than any mountain elevation on our own continent. Four or five Alleghanies would have to be piled on each other, and on them added Frémont's Peak, before their point would show itself above the surface. Between the Azores and the mouth of the Tagus this decreases to about three miles.' "

"But where was that 'submarine mountain' that there was so much said about in 1865 ? " I asked. "It was somewhere along the telegraphic plateau; because I remember that a great many persons argued, that, as there was a cliff two or three thousand feet high over which the cable had to hang, the motion of the water would soon wear it off."

"I recollect something about that," said Kit.

"Yes, there was such talk," replied Raed. "Lieut. Berryman found it. It occurs about two hundred miles west of Ireland. While making soundings for the telegraph, Capt. Dayman spoke of it as follows : 'In 14° 48' west we have five hundred and fifty fathoms' rock-bottom, and in 15° 6' west we have seventeen hun dred and fifty fathoms' ooze. This is the greatest dip in the whole ocean. In little more than ten miles of dis tance a change of depth occurs, amounting to seventy-two hundred feet.'

"And here, in the Report of the Second Telegraph Expedition of 1858, while they were laying the cable, it is recorded, that, 'about five o'clock in the evening, the steep submarine mountain which divides the tele-graphic plateau from the Irish coast was reached; and the sudden shallowing of the water had a very marked effect on the cable, causing the strain on and the speed of it to lessen every minute. A great deal of slack was paid out to allow for inequalities which might exist, though undiscovered by the sounding-line.' Afterwards the British admiralty had Capt. Hoskins sound more carefully; and the result showed, that, although there was a considerable descent, the idea of there being a stu-pendous cliff was all a humbug. Here's a slip that I cut from an English paper, and put into this report shortly after the last cable was laid. I'll read it, if you say so."

"Go ahead!"

"Speaking of the telegraphic plateau, the writer says, 'The dangerous part of this course has hitherto been supposed to be the sudden dip, or bank, which occurs off the west coast of Ireland, and where the water was supposed to deepen in the course of a few miles from about three hundred fathoms to nearly two thousand. Such a rapid descent has naturally been regarded with alarm by telegraphic engineers; and this alarm has led to a most careful sounding-survey of the whole supposed bank by Capt. Dayman, acting under the instructions of the admiralty. The result of this shows that the supposed precipitous bank, or submarine cliff, is a gradual slope of nearly sixty miles. Over this long slope, the difference between its greatest height and greatest depth is only eighty-seven hundred and sixty feet: so that the average incline is, in round numbers, about a hundred and forty-five feet per mile. A good gradient on a railway is now generally considered to be one in a hundred feet, or about fifty-three in a mile: so that the incline on this supposed bank is only about three times that of an ordinary railway. In fact, as far as soundings can demonstrate any thing, there are few slopes in the bed of the Atlantic as steep as that of Holborn Hill. In no part is the bottom rocky; and with the exception of a few miles, which are shingly, only ooze, mud, or sand, is to be found.' "

"It is said that these soundings show that the ooze on the ocean-bottom is just as soft and light as it is on the bottom of a mill-pond, despite all the immense weight of water resting on it," observed Kit. "That's a joke on the old philosophers, who used to say that the very water itself, at the depth of a mile, was solid as a rock from the vast pressure. I suppose the cable lies embedded deep in this ooze. How easy it would be for some big shark to bite it asunder!"

" There don't many fish go lown so deep in the sea
as the cable lies buried," replied Raed. "There is more
danger from the anchors of fishing-smacks off Newfoundland. It is reported, that, last season, a fishing-schooner
raised the cable on the fluke of her anchor to the top of
the water. They got it off, and let it drop back *as easy
as they could.*"

" That will do for a fisherman's story," said Capt.
Mazard, who had just come down. "I should sooner
think it was some old hawser lying on the bottom that
they pulled up."

" So should I," said Kit. "I don't believe they could
have raised the cable : it would have been too heavy
for them. When they grappled for the lost cable on
' The Great Eastern,' they had to connect the windlass
with the engine to pull it up ; and they broke any
amount of iron cable raising it."

" I should think, that, after a while, the water would
soak through the outside covering of hemp and gutta-
percha, and come to the wire inside the cable," said
Wash. "That would kill it, as I understand it. The
electricity would then go into the water."

" It would injure it," Raed replied. "But I believe
a naked copper wire will carry the electricity through
water without any coating. Isn't that so, Kit?"

" Yes. When they were laying the cable, they experi-
mented on that. They stripped a *foot* of the copper
wire bare, and let it down into the water, and sent the
signals through the bare wire almost as well as before."

" But there isn't much danger of the water wetting
through the coating," said Raed. "It is very impervious."

" You've got a history of the telegraph there, Wash," remarked Kit. " Read how the cable was made."

Wash turned over the book, and read : " The inside, or core, of the cable, consists not of one single straight copper wire, but of seven wires of copper of the best quality, twisted round each other spirally, and capable of undergoing great tension without injury. This conductor is then enveloped in three separate coverings of gutta-percha of the best quality, forming the core of the cable, round which tarred hemp is wrapped, and over this the outside covering, consisting of eighteen strands of the best quality of iron wire, each strand composed of seven distinct wires, twisted spirally, in the most approved manner, by machinery specially adapted to the purpose."

" 'That was the old cable," said Raed, — " the one that gave its last kick in 1858. The cable of 1866 is better than that. Turn along to that. You'll find it farther over."

" You're right ! " exclaimed Wash. " Here it tells about the last one : ' First, the central copper wire which was the spinal cord, the nerve along which the lightning was to run, was nearly three times larger than before. The old conductor was a strand consisting of seven fine wires, six laid round one, and weighed only a hundred and seven pounds to the mile. The new was composed of the same number of wires, but weighing three hundred pounds to the mile. This was made of the finest copper that could be obtained in the world, making a perfect conductor. To secure insulation, this condu tor was first embedded for solidity in Chatterton's compound, a preparation impervious to water, and then covered with four layers of gutta-percha, which were laid on alternately with four thin layers of Chatterton's compound. The old cable had but three coatings of gutta-percha, with nothing between. Its entire insulation weighed but two hundred and sixty-one pounds to the mile, while that of the new weighed four hundred pounds.

"'But a conductor ever so perfect, with insulation complete, was useless without proper external protection to guard it against the dangers which must attend the long and difficult process of laying it across the ocean. The old cable had broken a number of times. The new must be made stronger. To this end it was incased with ten solid wires of the best iron, or rather of a soft steel, like that used by Whitworth for his cannon. This made the cable much heavier than before. The old cable weighed but twenty hundred-weight to the mile, while the new one reached thirty-five hundred-weight and three-quarters. But mere size and weight were nothing, except as they indicated increased strength. This was secured, not only by the larger iron wires, but by a further coating of rope. Each wire was surrounded separately with five strands of Manila-yarn, saturated with a preservative compound, and the whole laid spirally round the core; which latter was padded with ordinary hemp, saturated with the same preservative mixture. This rope-covering was important in several respects. It kept the wires from coming in contact with the salt water, by which they might be corroded; and, while 't added greatly to the strength of the cable,

it gave it also its own flexibility: so that, while it had
the strength of an iron chain, it had also the lightness
and flexibility of a common ship's-rope. This union of
two qualities was all-important. The great problem had
been to combine strength with flexibility. Mere dead
weight was an objection. The new cable, though nearly
twice as heavy as the old *in air*, when immersed in
water weighed but a trifle more; so that it was really
much *lighter* in proportion to its size. This increased
lightness was a very important matter in laying the
cable, as it caused it to sink slowly. The old cable,
though smaller, was heavy almost as a rod of iron; so
that, as it ran out, it dropped at an angle which exposed
it to great danger in case of a sudden lurch of the ship.
Thus, in 1857, it was broken by the stern of "The
Niagara" being thrown up on a wave just as the brakes
were shut down. Now, the cable, being partially
buoyed by the rope, would float out to a great distance from the ship, and sink down slowly in the deep
waters.' "

"But even that was not the cable which now spans
the ocean," said Kit. "The one described there was
the one lost from 'The Great Eastern' in 1865, — the
one they fished for so long. The 1866 cable is an
improvement on that even. You look along a little
farther, Wash, to where they laid the 1866 cable: you'll
find it."

"Oh, yes! here it is: 'In the cable to be made for
the new line there was but little change from that of the
last year, which had proved nearly perfect. Science,
however, aided by experience, was constantly devising

some improvement. So now, while the general form and size were retained, a slight change in the outer covering was found to make the cable both lighter and stronger. The iron wires were *galvanized*, which secured them perfectly from rust or corrosion by salt water. Thus protected, they could dispense with th preservative mixture of the former year. This left the cable much cleaner and whiter. Instead of its black coat, it had the fresh, bright appearance of new rope. It had another advantage. As the tarry coating was sticky, slight fragments of wire might adhere to it, and do injury, — a danger to which the new cable was not exposed. At the same time, galvanizing the wires gave them greater ductility; so that, in the case of a heavy strain, the cable would *stretch* longer without breaking. By this alteration it was rendered more than four hundred-weight lighter per mile, and would bear a strain of nearly half a ton more than the one laid the year before.'"

"But this last one was laid from 'The Great Eastern,' was it not?" I asked.

"Yes. 'The Great Eastern' is the champion cable-layer," said Kit; "but that's about all she is good for. It's a grand big ship, though."

"How many tons burden?" Wash inquired.

"From twenty-eight thousand to thirty thousand, including her engines," Raed replied.

"Well, how many without them? How much cargo can she carry?"

"Fifteen thousand tons, I think it is. That includes her coal, though. For a voyage across the Atlantic, she has to ship from six thousand to eight thousand tons of coal. That's one reason why it doesn't pay to run her for freight or passengers : it costs too much. She's too big for the public demand, at present."

"Do you recollect how much the cable weighed ? " Kit asked.

"Not far from four thousand tons," said Raed ; "and the iron tanks that held it weighed a thousand tons more. If it had not been for 'The Great Eastern,' we might not have had a cable down yet : so she has done the world good service if she never has another job."

"Whom do you think the honor of starting the Atlantic Cable belongs to ? " Wash asked.

"Mr. Cyrus M. Field, to be sure," replied Kit.

"Some say Prof. Morse ought to have as much praise as Field," said Wash. "He was the inventor of the telegraph."

"Oh, yes! and, as such, the world honors him," replied Kit. "But he wasn't the man who put the Atlantic Cable through. Field was the man who did that, aided by the London capitalists."

"And the New-York capitalists," Raed added.

"Yes ; the New-Yorkers did something : but it was those London men who really pushed the thing through, after all. No use to try to hide that fact."

"The United States furnished the man who started the thing," Raed interrupted, "and who stuck to the idea, and never gave up till the cable was laid. Brains go before money. The honor is due us, rather than to England, I say."

"Of course ; but then brains can't work without

money to back them up. If it hadn't been for English
capital, there wouldn't be any Atlantic Cable to-day."

"If it hadn't been for Mr. Field, there wouldn't be
any Atlantic Cable to-day," retorted Raed. · "So there
you are again."

"To be sure. Let's divide the honor, then, between
the two countries."

"Yes; but not equally. America ought to have the
larger share," said Raed.

"I think half and half would be fair," replied Kit.
"Not that I want to deprive Mr. Field of any of the
honor so justly due him. I believe him to be the man
of all others most worthy of the imitation of all American young men, for the indefatigable energy and perseverance with which he stuck to the Atlantic-Cable
scheme. The failures and discouragements which attended that enterprise were greater and more disheart-
ening than those of any other I ever heard of or read
about. For twelve years he labored, often almost single-
handed and alone. After the failure of the 1858 cable,
he was ridiculed, abused, and suspected of the most
dishonorable motives. Most men would have given it
up in disgust, and let it gone to ruin. Served the public about right for its meanness too. But Field didn't
give it up. Then the Rebellion came on : that served
to stop every thing for a while. He still clung to the
idea. Why, that man has crossed the ocean something
like fifty times, I think, trying to push on the cable.
More than that, he embarked almost his whole fortune
in it, not from any hope of doubling his money or any
thing of that sort, but out of pure public spirit, and love

for the great enterprise.   Detract from that man's
fame! — no, sir! The man who tries to do that is a
mere blackguard."

"That's so."

" And that is why I say," continued Kit, " that there
isn't a man in this country so worthy to be taken as a
model of energy, perseverance, and real true philan-
thropy, as Cyrus M. Field."

"How about Mr. Frederick Gisborne?" I queried.
" Some say he's the man who first projected the Atlantic Cable."

" Mr. Gisborne is the man who first proposed a telegraph from St. John's, Newfoundland, to Halifax," said
Kit. " He worked hard to accomplish that, and is de-
serving of a good deal of praise. But there's no evidence that he ever dreamed of a cable across the
Atlantic. Some have said so since, I know; but I
never heard that he claimed it himself."

" Dinnah all ready, sah!" exclaimed Palmleaf,
thrusting his head into the midst of our debate.

During the forenoon of the 30th we passed a large
iceberg about half a mile to the north-west, bearing
slowly southward. A wreath of thin fog hung over it.
Two or three seals lay on a projecting mass just above
the water-line, and a flock of gulls wheeled about its
lofty pinnacles. There is scarcely a grander sight in
nature than one of these enormous ice-masses drifting
grandly down from the poles towards the equator,
never to return. The sun is its sworn foe: its grave
is the warm waters of the tropics. What becomes of
the seals that cling to it on its southern march? To be

left in mid-ocean a thousand miles from land, with the
water three miles deep, must be an awkward predicament, even for a seal.

On the 1st of June we had the company of a large
whale for several hours. We saw him blowing quite
early in the morning, several miles to the north. Gradually we came up with him. Immediately on perceiv-
ing the schooner, he came rushing down across our
bows. At first we thought he was going to butt us;
but his movements soon showed that he had no hostile
design. I think he mistook the hull of "The Curlew"
for another whale : indeed, he was nearly as long as
the schooner. After moving around us several times, he
fell in our wake; and, although we were sailing eight or
nine knots an hour at the time, the huge creature kept
following us from seven, A.M., till a quarter of eleven.
Wash wanted to fire at him with the big rifle;* but
Capt. Mazard said that he was too large a customer to
provoke. He might run his head against our side,
if angered.

The 2d, 3d, and 5th of the month were windy and
cold, with an occasional scud and sleet. We spent the
time reading up on Icelandic matters. We had Lord
Dufferin's "Letters from High Latitudes," Mackenzie's
"Travels in Iceland," Metcalfe's "Oxonian in Iceland,"
and several others.

* The " famous weapon " used during our last season's cruise

2

CHAPTER II.

ON the morning of the 9th we were in latitude 63° 43', longitude 26° 7' west of Greenwich; wind north-west; weather murky, cold, with dark driving clouds. It seemed as if we were indeed approaching that grim boreal region whence the storm-winds come.

"We ought to sight land by noon," remarked the captain.

We got our glasses, and looked off to the north-east. A dark bank of clouds rested down on the ocean. Between us and them the black waves with their foaming white crests heaved and tossed restlessly. Scattered here and there, patches of ice flecked the sea. Off to the north-west, toward Greenland, a long white ice-field lay along the horizon, extending far off under the dark cloud-masses.

By eleven o'clock the sky cleared, exposing a cold deep-blue horizon, with one white knob low in the ocean, just east-north-east by the compass.

18

"Is that an iceberg?" Kit queried.

"That's either an iceberg or *Iceland*," replied the captain: "I can't tell which, yet."

"Do you suppose their mountains are snow-clad at this time of year?" Wash asked.

"The high peaks are all the year round, save when the volcanic heat melts it off," said Raed.

We went down to dinner, and after that spread out our maps and charts, and compared them with our present position at sea, to make out what peak the snowy knob we had discovered could be.

"For it is certainly a mountain," said Kit, who had gone up after dinner to take another look at it with his glass. "It begins to loom up considerably; yet I can see that it is still a long way off. No iceberg would loom so high at so great a distance. Farther south, too, there are several other white points. It's Iceland: no doubt about it."

"Then that peak must be the Snæfels Jokul," Raed remarked.

"Why do they call it Snæfels Jokul?" Wash asked. "*Snæ* means 'snow;' *fell* means 'mountain:' together they mean 'snow-mountain.' But *jokul* mean. 'mountain' too."

"But there is a difference in the meaning of the words *fell* and *jokul*," said Raed. "*Fell* is a mountain which is not covered with snow in the summer, while *jokul* is a high peak or ridge clad with eternal ice."

"The proper name of this peak is Snæfel alone, I think," Kit remarked. "On some of the maps it is called Snæfel simply."

We went on deck for another look at it.

"If that is really the Snæfel," said Wash, "Reyk-javik, our port, must be east, or perhaps a point or two south of east. But we are heading direct for the mountain now, captain."

"Yes, sir : in an hour we will change our course, — as soon as we are sure it is the Snæfel."

As the afternoon passed, the snowy peak rose higher and higher : we were sailing fully eight knots per hour. To the north-east of it, several other peaks began to show; while to the south a long line of purple summits rose gradually out of the ocean. The Snæfel now seemed to stand far out towards us. This circumstance left no doubt that we were really approaching the entrance of the Faxa Fiord, and that the long purple arm extending out to the south-east was Reykjaines, - - the "Smoking Cape."

At four o'clock we changed our course to east-south-east, and stood down toward Reykjavik. We had brought our map on deck. To the south-east a high peak rose far inland.

"That must be the fiery Hecla," observed Kit. "Wish it would take a notion to *erupt* while we are up here!"

"And there are three other peaks inland just about east from us," remarked Raed. "Judging from their position, as compared with our map, they are the Eyrik Jokul, the Ok, and the Long Jokul."

At eight o'clock the sun was still high in the north. Snæfel was almost in line with it from the deck. To the south-east, a huge rock, many miles (ten) from the

black lava cape, rose abruptly from the ocean. Its top
was white as snow; its perpendicular sides were dark as
obsidian.

"That must be the famous Meal-sack," said Wash.

We had read of it. With a glass we could see the
waves breaking at its base, sending the foam far up its
sides. No one has ever landed on it. No boat dares to
come up near it. Its height has been estimated at two
hundred feet, and its diameter at a hundred and fifty
feet. Its white cap is doubtless from the excrement of
hundreds of sea-fowl which nest on it. Formerly it
was thought to be a breeding-place of that rare bird, the
great auk.

The wind held strong from the north-west. "The
Curlew" flew on. By ten o'clock we had got far up the
fiord. The coast about the head of it was now in plain
sight, — a rough, black shore, distant ten or a dozen miles.
The nearer we came, the rougher and blacker it looked.
The point to the south of us seemed to be a bare black
lava-ridge, rent and shattered in the most frightful manner. Not many years since, there was an irruption in
the sea here, during which these dark lavas were cast
out. The entire head of the fiord in front of us seemed
serrated with jagged headlands and points, between
which narrow inlets ran back as far as we could see.
There were numerous small islands, — so many dark,
dangerous-looking ledges. In fact, it was, as Capt.
Mazard expressed it, about the "wickedest-looking coast"
he was ever on.

But where was Reykjavik? We began to ask the
question ; for we were now within four or five miles of

the shore, and had not yet espied the port. Fron the general correspondence of the coast, the peak of Snæfel, and the Meal-sack, with our chart and map, we felt sure we were in the Faxa Fiord; and somewhere in the southern part of this great fiord, or bay, was Reykjavik The chart indicated a passage between a point and an island as the entrance to the harbor; but so many friths between islands and points showed themsel-es all along, that we were in considerable doubt. Presently we espied a fishing-smack dart out of an inlet a mile to the northward, and come running down along the coast. As we came up, the smack passed across our bows at the distance of half a mile or more, and, bearing southward, went in between two small islets, and disappeared.

"Going into port, most likely, at this time o' day," said the captain. "Guess we'll make bold to follow."

The helm was set a-starboard, and the schooner headed for the passage. Just outside the islands, we met a brig coming out. This confirmed us in our previous opinion; and, on passing in between the islands, lo! the harbor, with some half a dozen craft of various sizes, disclosed itself a couple of miles to the south-east. Half an hour more, and we were in the roadstead, and had dropped our anchor some two hundred yards below a little black jetty, which was the only approach to a wharf anywhere in sight along the water-frontage. A hundred yards to the left of us lay a small ship-of-war with the French flag flying. We could see her officers looking at us rather curiously, and immediately ran up our bonny new flag. It's not often that they see the stars and stripes here, I fancy. It was now past eleven; but the sun was

still shining, and showed a fair hand-breadth above the
snowy mountain-tops to the north. We looked off to
the town. Every thing was quiet. Three or four men
were getting ashore from the smack in their boat:
otherwise there was no stir, no noise, save the dismal
howls of a couple of dogs that seemed to be practising a
rancorous duet.

"Folks are probably abed," Kit remarked. "I suppose they have to sleep, if the sun doesn't set."

"But is it possible that this is really Reykjavik, the
capital of Iceland, — this contemptible-looking little
collection of paddy shanties!" exclaimed Wash. "Why,
the meanest fishing-village on the New-England coast is
a gay metropolis beside this. I do believe we must be
mistaken. Haven't we got into the wrong port?"

"Oh! this is Reykjavik fast enough, isn't it, captain?"
Raed inquired.

"I don't think there can be any doubt of it," said
Capt. Mazard.

"But I can't believe it — hardly!" Wash cried.

"Those Frenchmen seem to be eying us quite curiously," remarked Kit. "Might try your bad French on
them, Wash."

"Yes. Ask them what place this is," said Raed.

Wash cleared his throat.

"Excusez moi, messieurs: cette ville, est elle Reyk
javik?" he shouted.

There was a moment of doubtful hesitancy, then a
very politely modulated —

"Excusez moi, monsieur: seriez-vous assez bon pour
repeter ce que vous avez dit?"

"He means, say it over again," said Kit. "He didn't understand you."

"Cette ville, est elle Reykjavik!" repeated Wash.

"Oui, oui, monsieur: *cette ville* est Reckjahveek," was the reply.

"We heard them laughing, whether at Wash's queer French, or at our having got to a place we didn't know the name of, was uncertain.

"Bet you they are jolly fellows!" Kit exclaimed. "Should like to get acquainted with them a little."

"Well, this settles the point then," Raed remarked. "Yonder little hamlet is the largest town of Iceland. Why, there can't be over a hundred shanties. There aren't half a dozen buildings that really deserve the name of 'house:' the rest are mere hovels. This, then, is the capital of an island larger a good deal than the State of Maine."

Maine has thirty-one thousand seven hundred and sixty square miles, and Iceland has thirty-seven thousand," remarked Kit.

"But the population of Iceland is not over seventy thousand in all," said Wash. "Maine has nearly ten times that number."

"And yet this is the country, and yonder hovel-dwellers are the descendants, of the men who first discovered America," said Raed musingly. "As early as the tenth century, long before ever Columbus was born, or dreamed of crossing the Atlantic, these hardy Icelanders had followed down the North-American coast as far as Massachusetts, and established colonies. The famous 'round-tower' at Newport is thought to be the work of these early navigators."

"But Longfellow, in his poem, 'The Skeleton in Armor,' fancies that it is the work of a Norwegian viking," I observed.

"Oh, that's a mere poetic license!" cried Kit. "It's far more likely to have been the Icelanders."

"But the Iceland people are the descendants of the Norwegians, are they not?" Wash asked.

"Partly," said Raed; "though the Irish are said to have had a hand in settling Iceland, and also the English. I have no doubt that the Norse vikings were the first to set foot in Iceland. One of the sagas says Naddothr, a Norwegian pirate, first discovered Iceland, — in the year 860. At that time the island was uninhabited. So dreary did it look even to his northern eyes, that he called it Snja-land, and forthwith sailed away."

"It is said that Columbus had made a voyage to Iceland before he discovered America," remarked Kit. "Many think he heard of the Western continent from the Iceland folks."

"Yet he thought it was the other shore of the Eastern continent all the while," Raed observed. "No doubt he heard traditions of these Norse discoveries; though it is doubtful if he got any very definite information, else he never would have set out to cross the Atlantic at its very widest point, as he did."

"I have no doubt he did get a great many ideas of the 'Western land' from this people," said Kit; "for, at that time, the Icelanders had colonies on Greenland and on Labrador."

The sun was just touching the mountain-peaks of this strange land of ice and fire.

"Let's turn in, and get rested for to-morrow," Raed advised.

The watch was assigned to Smith and Trevers. We went to our bunks, and slept soundly till after four, when the shouting and *heave-ho-ing* of our French neighbors aroused us. They were weighing anchor, — making great din about it too. Strange what a difference there is between an English and a French ship in this respect! An Englishman will weigh anchor and leave port without a sound save the noise of her blocks, capstan, and the officers' commands. On a Frenchman there is always a great *to-do* and jabbering. Monsieur must *exclaim*, or he couldn't do a thing. This difference is, I presume, due partly to the more excitable temperament of the French, and partly to the stricter discipline on an English ship.

"Sorry they are going off," Kit muttered. "I hoped they would stay a while."

"They are going up the west coast, I think it likely," said Capt. Mazard. "The French have a great many fishing-vessels about this island in the summer time. This brig-of-war is up here to protect them, probably, and look out for their interests."

After breakfast we let down our boat; and we four boys, with the captain and Weymouth and Hobbs, rowed up to the little rickety jetty, and landed for the first time on Icelandic soil. Hobbs and Weymouth remained with the boat. Several fishermen with pipes in their mouths were lounging about; but, as they said nothing to us, we did not accost them. A little up from the shore, at the beginning of a motley street, we espied a board house, which from a notice in Danish, of which we could make out a word or two, and the picture of a steamship in rapid career, we concluded to be the office of the Copenhagen steamer.

"Let's call and inquire for the governor's house," proposed Raed. "There's a Danish governor resident at Reykjavik, you know. He's the man we want to see."

But the steamer-office was closed.

"Too early, perhaps," said Wash.

We strolled along up the street, between a row of one-story board houses on one side, and an irregular collection of hovels, some of them built of rough stones, with turf roofs, on the other. Turning off to the right, we at length came out beside a plain little church with a queer, box-shaped steeple. A muddy street led off past it to the left. Leaning against a hovel on the corner was a man, with his fur cap over his eyes.

"Hollo, friend!" exclaimed Capt. Mazard, walking up to him. "Can you tell us where the governor lives?"

The man stared.

"*Nicker forstay,*" he muttered, — something which sounded like that, — he didn't understand.

"Try him with the word *amtman,*" said Raed. "That means 'governor' in Danish, I believe."

"*Amtman?*" cried the captain, swinging his arm over the place.

"*Yawh!*" (a sentence perfectly unintelligible,) — pointing to a large squat stone house two or three hundred yards farther on.

We thanked the fellow with bows, and went on.

The residence of the governor is a one-story house
of stone and mortar, originally designed for the State-
prison of the island, it is said; but either from the
number of convicts getting too small, or the State too
poor to support a prison, it was metamorphosed into a
governor's house.   It enjoys the reputation of being the
most splendid mansion in the country; and, from what
we afterwards saw, I dare say it is.   A stone wall
separates it from the street on one end.   In front is a
lawn sloping down to a sort of common, or square, from
which it is set off by a wooden fence.   Such a fence is,
I dare say, something of a luxury in Iceland.   The
wood is imported from Denmark.   Racd and the captain went up to call on his Excellency.   The rest of us
continued our ramble about town.

Off to the north-east of the governor's house, and
standing a little apart, is a high (for Iceland) building, —
a Latin school, we learned.   The Reykjaviks call it a
"university."   Here it is that their clergy and other
professional great men are run through the Latin and
Greek mill.   A four-years' course of dead languages is
absurd enough in America; but somehow it seems
doubly so amid the squalor and poverty of Iceland.

"When will the world get enough of this monkish
study of Latin and Greek?" Kit exclaimed as we
stood looking at the university.

"That is one of the legacies the Catholic religion has
bequeathed us," laughed Wash, — "that every son of
Christian mother born must have two big lexicons slung
about his neck from his tenth till his twentieth year."

"Thank our stars we have broken loose from them !"

cried Kit. "I wish every boy in the United States had the courage to do so. Six months of Latin is enough. We Americans have got to give the world a model in educational matters as well as in many other things. The old idea of cramping a boy down in some little stived-up town to spend four and eight years in getting his 'education' is monstrous. Why, the whole world is the book for him to study and learn from. He must go about it, studying as he goes."

On the hill a little back of the governor's house is a windmill, where the daily bread-corn of the town is ground; though just what this bread-corn consists of we have yet to learn. A few watery potatoes, with straggling patches of oats and wild corn (not maize), constitute the agricultural products of the country. The climate is far too severe, and the season too short, for wheat or Indian corn.

On going past the church to the left, we came out on the shore of a pond, the outlet of which, a small brook, runs through the common in front of the governor's house, and thence down into the sea. A reedy, boggy, bleak-looking tarn, it seemed to us. Several grebes and a fine merganser started up as we came out to the water's edge. The water gave an icy chill to our fingers. We saw ice on the farther side. Coming back past the church, we heard a great shouting and hallooing in the direction of the jetty.

"Some sort of a row!" cried Wash. "Our boat with the sailors is down there too!"

We hurried forward, and, passing the steamer-officer, caught sight of a crowd about the jetty. Over their

heads could be seen a couple of blue caps, with now and
then a brawny arm flashing up.

"It's Hobbs and Weymouth!" muttered Kit.
"Fighting the whole crowd. I wonder what the row's
about."

"Push in!" cried Wash. "We must stand by
them."

The frowzy, broad-faced mob stared in our faces as we
hustled them aside. Arriving on the jetty, we found
the two sailors "squared off," with sleeves rolled up,
warding off the grabs of two men, whom, from the
badges on their coats, we took to be policemen. They
fell back as we pushed past them, talking very loud and
authoritatively.

"What's the matter?" Kit demanded. "What's
the fuss about?"

Hobbs grinned.

"You see, sir," explained Weymouth, a good deal
out of breath, "we were sitting here on the wharf,
when along came as pretty a lass as ever I set eyes on.
She went to one of those boats out there on the shingle,
and began to take out an armful of fish. Well, Hobbs
he watches her a while, and then strolls along where she
was, and told her she was much too pretty a girl to be
handling them slimy fellows, and asked her if he
shouldn't lug 'em for her. I don't know as she understood him; but she let him take the fish, and then piled
more of them on his arms. When they had got an
armful, she led off, and Hobbs started to follow with the
fish. I had noticed a chap standing up there by that
shanty yonder. The minute they started away from the

boat, this chap he came along and stepped in front of them, and began to swing his lip, and by and by gave the girl a slap on the mug. At that Hobbs dropped the armful of fish, and gave the chap a mate to his slap, right across the eyes. Then the chap went for him; and Hobbs knocked him over into the mud. He got up and ran off a bit, and began to jaw, and shake his fist. Upon that Hobbs puts his thumb to his nose, and *twinkled* his fingers at him (illustrating it). Just then one of these fellows (pointing to the men with the badges) comes along. The chap entered complaint to him, and then went off after the other fellow here; and together they tried to fasten on to Hobbs. But he backed on to the jetty with me, and we've kept 'em off."

" In other words," exclaimed Kit, " you are resisting the officers, the police ! "

" The police ! " cried Weymouth, considerably sobered by the thought, and looking them over with serious attention. " Well, I'll be blamed if they aren't the rummest-looking policemen I ever clapped eyes on. I don't believe they amount to much."

Kit turned to the officers, who did seem remarkably patient under the contumely with which the Reykjavik law had been treated in their persons, and, pointing off in the direction of the governor's house, said, *"Amtman,"* at the same time taking a handful of gold dollars from his waistcoat-pocket. The worthy guardians of the Icelandic peace hesitated a little, not understanding, probably, on what terms we stood with the *amtman.* Seeing this, Wash gave the sailors the nod to get into the boat. They did so in a twinkling, and shoved off

Making our way off the jetty through the crowd, we caught sight of a fellow with a bloody nose, which he was mopping with a very dirty handkerchief. Near him stood a young woman of a fresh, rosy complexion, looking demurely downcast.

"There's the fair *casus belli,*" laughed Kit.

"And there's the chap *wot* got hit in the mug,' added Wash.

We approached him, and, by sundry serious shakes and nods, tried to give him to know that we regretted the injuries he had suffered. When he had checked the "claret" from his nose, Kit, with many pitying shakes, slipped into his hand three or four dollars; seeing which, the crowd began to nod approvingly to each other. We walked off. Truly money will heal most wounds. Thus ended Hobbs's adventure. Going back toward the church, we met Raed and the captain coming away from the "stone house."

"Well, what said his Honor?" Wash asked.

"He gives us a hearty welcome to Iceland!" cried Raed. "Pledged us health in a glass of port and no end of corn-brandy. Nearly upset the captain!"

"And no wonder!" cried Capt. Mazard; "for I had to drink for both of us. Was afraid his Excellency would take offence if I refused. That comes of being out with a temperance man."

"He gave us lots of information about travelling in land," Raed went on. "No roads, no carriages : every body travels on horseback. We have got to buy our horses too. Don't *let* horses here, it seems. Must have two apiece."

"Whew!" exclaimed Kit: "that will cost us something."

"But he says one can purchase a good horse here for twenty-five dollars," replied Raed.

"How about a guide?" Wash inquired.

"For a guide, he directed us to apply at the hotel. Didn't see the hotel, did you? Well, they have one. It's that two-story wooden building up beyond the church. We've been up there. The landlord speaks English, — after a fashion. So does the governor, for that matter. Landlord promised us to notify half a dozen men who sometimes act as guides. They are to come to the hotel at three o'clock this afternoon. We must be there then to take our pick of them. So let's go aboard and have dinner, and get rested."

"Raed is vastly taken with the governor's daughter," observed the captain.

"Of course I am!" exclaimed Raed. "So would anybody be. She is a very pretty young lady (anybody can see that); and she is really refined and accomplished, to say nothing of her beauty. She does the honors of the house with an ease and grace that I didn't much expect to see in Iceland."

"Bravo, Raed!" cried Kit. "Sounds as if he might be son-in-law to an *amtman*, yet; doesn't it?"

"I am not so presuming as that, I assure you!" replied Raed.

We went down to the jetty, and signalled the schooner to send the boat; for Hobbs and Weymouth had discreetly betaken themselves aboard.

After dinner we read a while; then, going ashore

3

again, went up to the hotel. It at least resembled
hotels in this respect, that it had a bar whence corn-
brandy, beer, and other liquors, were being dispensed to
a whole roomful of smokers and men holding horns of
snuff, which they from time to time applied to their
noses.   These snuff-horns — which, by the by, are a sort
of national institution in Iceland — are much like a small
powder-horn.   When the Icelander wants to take a
pinch, — which is pretty often, — he sticks the little
end in his nose, and snuffs.   Such a whiff would split
the nose of any ordinary American with a paroxysm of
sneezing; but an Icelander's nose is of less sensitive
make.   When an Icelander does sneeze, however, he
does it with a vengeance, sending a cloud of snuff in
all directions.

Only three of the professional guides whom the
landlord was to notify had made their appearance.
They were sitting in a row on a bench, waiting our
pleasure.   One was a little thick-set fellow, with a
broad, flat face.   The landlord introduced him as Guthmundr Gissurson; that is, Guthmundr, the son of Gissur.
He had been all over Iceland several times, save the
regions to the south-east, where no one ever goes.   All
he asked for his services was one dollar (Danish money)
per day.   The second was a large, plainly-dressed man,
with a frank, honest face, named Zöga.   His price was
a dollar and a half per day; but he would furnish his
own horse, — quite an item where all the horses had to
be bought.   The third was a young fellow, quite tall,
and rather sprightly for an Icelander, who is generally
just the reverse of sprightly.   His name was Halgrim

Arnarson. He was the son of a farmer living two miles (Icelandic miles, each of which is equa. to five or six English miles) out of Reykjavik. He had already escorted two English tourists to the geysers. From them, and from divers Englishmen he had met in Reykjavik, he had gained enough of the language to understand us, and answer in brief. Zöga, on the contrary, had quite a command of English. He had taken many parties to the geysers, and always given satisfaction, — so the landlord told us confidentially. Young Arnarson, however, suited us best. He could, moreover, furnish us with six horses from his father's farm ; and, if they were returned *sound*, we were to pay him but six dollars apiece for their services. His price was one dollar per day and board. We engaged him.

For carrying our tent, kettle, &c., together with what provisions we should need to take along from the schooner, we learned that six horses would be necessary, — a statement that startled us not a little, till we came to see the horses. Then, for our own riding, each of us was obliged to have two, in order to shift our saddles once in every three or four hours. We had intended to take Palmleaf along to do the cooking for the party ; but this great expenditure for horses staggered us not a little. At this rate, we should have to pay five or six hundred dollars for horse-flesh at the very outset. Kit at once proposed that we should leave Palmleaf on board, and thus save at least two horses. Capt. Mazard could not go: he did not deem it pr ident to leave "The Curlew" with the sailors. There was no knowing what scrape half a dozen young tars might get into if left to themselves. We at length concluded to take Weymouth only. With the guide, there would then be six of us. The baggage was finally *argued* down to what five horses could carry. There would now be needed seventeen horses. Eleven of these must be bought outright. Halgrim agreed to notify the horse-dealing portion of the little town of our wants. We set seven o'clock that evening as the hour for them to bring on their nags, and went back on board the schooner. Had supper at half-past six, and shortly after went off to the jetty again. On nearing the hotel, we beheld not less than a hundred persons, nearly every one of whom held a horse, and some had two, — quite a chance for an extended choice certainly.

"But do look at the brutes!" Wash exclaimed. "Why, they are shaggy as lions, and nearer the size of billy-goats than horses."

"I've read that the Icelanders feed their horses on dried fish," cried Kit: "now I believe it."

What amazed us most, next to their extreme littleness, was their shagginess. Such manes and tails! Why, one of those tails would stuff a mattress! In a moment we were surrounded by the whole crowd, each man hanging to his halter, and crowding up to add his own voice to the din of unintelligible lingo which was raised at our approach.

"Well, of all the vicious-looking little monsters, these beat every thing!" Raed exclaimed. "Look as if they would make nothing of snapping off a fellow's arms or his feet; mistake 'em for a dry fish. Let's get out of this. I'm half afraid of 'em."

By dint of pushing and sharp dodging, we gained the door, where our guide stood coolly surveying the assemblage.

"Ready — buy — my sirs?" he remarked.

"It's no use for us to try to buy of them," groaned Raed. "Halgrim, you buy. Buy good ones, Halgrim — good ones."

"Yas, sirs," said Halgrim.

We got into the bar-room, and, going to the window, stood by to see the trading go on. Well, I suppose it didn't differ so very much from horse-traffic in other parts of the world; but together with the droll sound of their language, the twitching of halters, the biting, shrill squealing, and occasional wheeing-up, of the absurd little ponies, it amused us vastly. Halgrim examined the horses with great care and judgment. At first we were a little suspicious he might act with the jockeys, and either pay just what they had a mind to ask, or perhaps buy poor animals at regular prices to accommodate his friends; but he was honorable. It took him toward an hour to purchase the eleven. As fast as he bought one, he would hitch it to the fence to the right of the building. When the eleven were at last tied in a row, it was, to say the least, a singular spectacle. All the colors were represented, — brown, black, gray, calico, piebald! The bill was four hundred and forty dollars, Danish currency. And here a difficulty arose. When we came to produce our American gold in payment, — we had brought along three hundred dollars from "The Curlew," — the Reykjavikers shook their heads. They didn't know any thing about such money. Fortunately, however, there is a Danish exchange-broker in the town. Thither the landlord directed us. For our three hundred dollars American currency we got from this gentleman five hundred and twenty-five rix-dollars. He took a heavy percentage, as we saw on looking up the regular exchange value. The rix-dollar is worth only fifty-two cents of our money: fifty-two cents five mills, I believe it is.

Leaving the horses in Halgrim's care, with orders to have them fed and stabled, and also directions to hire or buy six riding-saddles and five pack-saddles for our provisions, tent, &c., we went on board for the night. Thus closed our first day's experience at Reykjavik.

CHAPTER III.

WE were early astir next morning. One hundred pounds of ship-biscuit were brought up from our stores, fifty pounds of cheese, ten of sugar, five of coffee roasted and ground, and two pounds of tea, with salt, pepper, &c. We also took a quantity of macaroni, and several tin cases of preserved meat for soups. Then there were compass, great-coats, one of our rifles, and a shot-gun, fish-hooks for Icelandic trout and char, and a portfolio with sheets of drawing paper and pencils for sketching ; for, during the winter, Wash and Kit had taken lessons in pencil-drawing.

By eight o'clock we had got through breakfast and had our luggage stowed in the boat. It was let down, and we took our places.

" Good luck, boys! " cried Capt. Mazard. " Success and pleasure attend ye ! "

" Wish you were going with us ! " exclaimed Raed.

39

"Never mind : I shall contrive to amuse myself. I may call on the governor's daughter. What say to that, my boy ? "

"Ahem! Well, I hope you will take care of my interests there," replied Raed.

"That I will; also of my own. She's a very fine young lady. I shall try to make myself agreeable. Good-by ! "

" Raed, the captain's rather got the advantage of you just now," laughed Kit as we pulled in toward the shore.

"Oh ! I'm not afraid of the captain," said Raed.

Smith and Bonney rowed us up to the jetty. Three fishermen were hired to take our luggage up to the hotel, where we found Halgrim, with his nags " all saddled, all bridled, all ready " for a start.

Tourists have often complained — justly, I have no doubt — of the stupidity and invincible laziness of Icelandic guides; but it is due young Arnarson to say that he was an exception to this rule. Raed thought he was as prompt and stirring as an average Yankee youth, who, in his opinion, combines all the excellences of all known races, — provided, always, he hails from Massachusetts; and the nearer Boston, the better. My cousin Wash is precisely of the same opinion. Kit, who comes from Northern Maine, holds rather broader views. I can get along with Kit very well; but for a fellow having the misfortune to be born in Georgia to fraternize with a couple of native Bostonians requires patience. Why, a fellow can't even express a sound opinion, unless he will acknowledge that he got it from

Elward Everett or Charles Sumner, — particularly the latter; but, if they don't change their minds about Sumner within the next five years, I am no prophet. That man has been kept in office, and flattered with public applause, till he really thinks he's a demi-god. It is a consolation to us Southerners, who have heard him fulminate so long, to know that all recorded demi-gods have ultimately come to grief. Sooner or later, these wonderful beings have uniformly developed a weak spot which floors them.

Halgrim was ready; so also was a little greasy-looking Dane, with a bill of eighty-seven rix-dollars for saddles and wooden boxes with locks for our pro-visions. These boxes are hooked on to the pack-sad-dles, on either side. While we were settling the bill, and effecting an exchange of another hundred dollars with the broker, Halgrim adjusted our luggage, and tied the five pack-horses with the six extra saddle-horses in line, the nose of one to the tail of the one next ahead: this was to prevent them from straying. Eleven of these absurd shaggy ponies in a line made about the queerest sight imaginable. To start this cavalcade without breaking the line seemed a rather nice job. Halgrim accomplished it, however. We then mounted our saddled ponies, and started off at a *lope,* our feet almost touching the ground. In any American city, I fancy we should have created a sensation. Once on a gallop, very little could be seen of the eleven forward ponies, save a wildly-drifting mass of hair, flying manes, and irrepressible tails streaming up high over the rout. The most distressful grunts resounded all along the line, accompanied by loud puffings and an occasional squeal, which, with the sharp *bookerty-book* of their iron-shod feet on the hard lava pebbles, made a din altogether ludicrous and ridiculous. Add to this our guide, with broad-brimmed, bell-crowned hat, closing up behind the animals, and brandishing a long whip, with loud shouts of "*Afram-yho!*" ("Go 'long!") and "*Hur-r-r-r-r, hur-r-r-r-r, hur-r-r-r-r!*" (equivalent to our "Hi, hi, hi!") and the reader will have a picture which he can complete by imagining our party in full career behind, putting on the whip; for each of us had been provided with a riding-whip with our saddles. On we went, full tilt, (and plague take the hind one!) past the governor's house, past the tall windmill, and out of town, with half a dozen curs yelping after us.

The sun was high up in the heavens: indeed, it had not been *down* at all that night. Weymouth, who had had the watch from midnight till one o'clock, told us, that, at half-past twelve, fully a third of the sun's disk had been visible over the white peaks to the northward, and that it had soon come into plain sight above them. But the air was chill, for all that. Fifteen or twenty miles to the eastward, a ridge of snow-clad peaks gave the country a wintry aspect, despite the bright sunlight. Every thing was so silent too, so still and voiceless, that a strange feeling of loneness crept over us as we scampered along. No song-birds enlivened the June morning with their carollings. There were no trees, no shrubbery beside the way, and very little grass. A couple of miles out of town, the road had dwindled to a mere trail. No fences nor walls enclosed it. The whole

country about was a common, — a bleak, black lava des-
ert, uncultivated and barren, with here and there a dull
dark pond, stagnant within its sedgy shores. The
gusts of chill wind from the icy jokuls gave a *shivery col-
oring* to the desolate landscape. We had not expected
to see any thing like tropical or even temperate scenery
in Iceland; yet I must confess to some disappointment
in this my first view of the country. It was more
cheerless and dreary than I had fancied it would be.
Knowing that the island boasted a hardy, honest peas-
antry, I had thought to find snug little cottages, surrounded by walled fields and green meadows. Alas for
this ideal picture! The cottages are mere huts, — cobble-
stone walls with turf roofs. The fields are sterile and
rough enough to appall even a New-Hampshire·farmer.
Barren moors, covered with rough fragments of lava;
bleak valleys, filled with cold morasses; dun-colored
jokuls, with black foot-hills rising to snowy peaks;
rapid rivers, foaming amid sharp black bowlders; fearful
cracks and yawning chasms in the vast lava-beds which
have been poured out over the country; hot springs and
steaming pools, — these are the features of Iceland scenery. Heaps of stones rudely piled up enable the traveller to keep the trail. Up hill and down hill, onward,
we went, Halgrim cracking his whip, and shouting.
It was marvellous how well the ponies made their way
at such a pace and over so rough a road. For an hour
or two, our chief business was holding on by the mane
and keeping up. Conversation was nowhere. All at
once, the leading pony bolted from the trail off between
two crags to the left; the second and third followed

him; but the fourth, either from better discipline, or
being under too great momentum, forged ahead. Snap
went the halter or the pony's trail! Forward went the
hinder eight in good order; but the three leaders galloped off at a great rate, and, disappearing among the
rocks, came out on the moor beyond, and ran like deer
toward the top of a long ridge half a mile off. A shout
of dismay arose from the whole party. All the bread
was packed in the boxes on the back of the first pony:
the second bore all the cheese and sugar. To see our
supplies vanishing in this way was demoralizing.

How far the vicious little beasts would have run, or
where they would eventually have halted with the
bread and cheese, nobody knows. Fortunately for us,
the hind one was the swiftest of the three. In the race
up the side-hill, he so far outran the forward two as to
whirl them both round. This manœuvre brought them
all three to a standstill; and while they were squealing,
biting, and kicking each other, in their attempts to get
loose, Halgrim headed them on one side; and, Kit and
Weymouth riding up on the other, the runaways were
secured, and driven back. The rest of us, meanwhile,
had overtaken and halted the other eight. The three
bolters were tied on behind, and the train proceeded.

About eight miles out of Reykjavik we met a pack-
train of seven ponies, tied together like our own, laden
with wool. It was attended by an old man with a very
long white beard, and a boy of thirteen or fourteen.

" *Vær-thu-sœl!* " cried the old man as he came off
opposite us, "*vær-thu-sœl!* " (" May you be blessed! ")

" *Vær-thu-sœl!* " responded Halgrim reverently; and,

dismounting, he fell on the old gentleman's neck, and
kissed him with a vast show of affection. The boy, a
dirty chunk of a fellow, was then embraced and kissed
in much the same way. They talked for some minutes.
Snuff was then taken from the old man's horn; and,
after another embrace, we rode on.

"Was that your father, Halgrim?" Kit asked.

"No, sirs! no father," replied the young Icelander
in some surprise.

We had thought it must be his father.

"Is he your uncle, or some old friend?" Wash inquired.

"Oh! I have him once before seen," remarked Halgrim.

"Do you always kiss your acquaintances like that,
Halgrim?" Raed asked.

"We our friends always kiss," was the grave reply.

"Gracious! What would they think of such a cus
tom as that at Boston?" Wash exclaimed.

"Do you kiss the girls — the *skën jumfrus* — in the
same way, Halgrim?" Kit inquired.

"Yas, sirs; we the girls always kiss," was the demure
reply.

"Why, that must be pretty rich, Halgrim!"

"How about the old women?" exclaimed Weymouth.

But the guide did not reply directly; and, seeing that
he did not like this last question, Raed changed the
subject.

At ten o'clock we halted to change our saddles to the
fresh horses, and let them all breathe for a quarter of an
hour.

Shortly after we came to a narrow bog, through which the ponies floundered heavily, throwing the mud and water about to such an extent as to thoroughly bespatter every thing. There were numerous sloughs of inky water and mire, into which the shaggy little nags plunged up to their backs. Nothing save the tightness of the pack-boxes saved the bread and cheese this time. Halgrim lashed the forward animals through with encouraging shouts of " *Gä, Gä, Gä!* " ("Go it!")

Then came our turn. Starting a rod or two back from the edge, we whipped up and went at it with a rush, *spatter-spludge.* Weymouth and Raed got across all right; Wash and I also succeeded in getting out pretty muddy, spitting vigorously to get the nasty water out of our mouths; but poor Kit was less fortunate, as a great splashing and outcry announced. Kit is a fellow who doesn't like to be the last man anywhere : but Fate was against him this time; for, in sheering to the left to get past Wash and I, he got his pony out of his depth, and came to an awkward pause, half buried in the mud, and dank, grassy tussocks.

"Say, Kit! — what are you up to there?" shouted Raed.

"Oh! I thought I'd let *him* stop and drink," said Kit.

"Wouldn't let him drink too much," laughed Wash. "He's *settling* fast."

"So I perceive," replies Kit quite coolly, drawing his legs up out of the muddy stirrups. Then, standing up on the saddle, he leaped to a tussock, thence to another, and so out to the bank.

How to get the horse out was a problem.

"What say to that, Halgrim?" Raed asked.

Halgrim looked dubious. The pony was some twenty or thirty yards from the firm ground, and nearly even with his back in the mire, which shook alarmingly as he plunged and floundered about.

"A genuine 'quaking bog,'" muttered Raed, — "such as they have in Ireland and Newfoundland."

"In Brazil they call such shaking masses *tremendals,* I've read," said Wash.

"One tremendous, trembling *tremendal!*" cried Kit. "How am I going to get my nag out of it? Can anybody suggest any thing?"

"Might twitch him out," said Raed, "if we had a rope."

"There's the guy-ropes of the tent!" exclaimed Wash.

"Just the thing."

Halgrim at once unpacked them, and we tied them together. Halgrim and Kit then pulled up their pant-legs, and, keeping on the tussocks, worked out, and made one end of the rope fast to the saddle. The Icelanders then took the pony encouragingly by the bridle; and the rest of us, getting as good a foothold as possible among the tussocks and on the shore, straightened the rope. "*Hurr-r-r-r, Hurr-r-r-r!*" shouted the guide. The pony sprang for life. We all surged at the rope, and, getting a start, had the wretched little brute out all standing. He trembled and shook as if he had acquired that motion of the bog. He was a singular-looking beast before he had gone through the slough; but now he was quite irresistible. All his vast luxuriance of hair was now loaded and dripping with mud; and, to relieve himself, he began a series of *shakes*, which kept everybody at proper distance. Kit managed to get the saddle off, and, after scraping it, put it on his second pony to give this one time to *dry*. We remounted; and thus ended our first adventure in the bogs of old Iceland. It had been a dirty scrape.

These Icelandic morasses are quite destitute of brush or bushes. A kind of long grass, springing up during the short summer, and matting down upon the mud during the winter, collects, in time, often in such quantities as to enable one to cross on it dry-shod. Many of the bogs are covered by huge tussocks, as large as a barrel standing upright, in countless numbers. One can sometimes cross on these tufts by jumping from one to another.

A little after noon we halted an hour for bread and cheese, and to let the horses graze on the side of a damp hill which offered a sparse growth of grass. One farm-hut, or *byre* as Halgrim called it, was in sight a quarter of a mile to the right. We did not go out to it. Milk and coffee of good quality can always be had of the cottagers, it is said, if one can stomach the filth and foul odors which pervade the huts.

We went on, and toward four o'clock began to ascend steadily toward the summit of a long lava-ridge forming the flank of a plateau of higher ground than the moors of our morning ride. A great deal of loose lava lay about, — angular, black fragments, and sharp, bristly ledges, which were fearfully suggestive of broken heads

as we rode over them at a gallop. An hour later we
turned in between two almost perpendicular cliffs, which
towered for sixty or a hundred feet above the trail, with
scarcely space for two to ride abreast between them.
These cliffs are black as soot, with knife-like edges, and
angles such as one only sees in a volcanic region. Passing between these rocks, we descended into a long,
narrow defile, walled in on either side by sheer black
precipices towering to the very clouds. The entrance
had a very sombre seeming.

" What place is this, Halgrim ? " Kit asked.

" The Almannajau, sirs." (Pronounced, All-mannah-
gee-ow).

" What, the famous old Almannajau, where the ancient
Iceland senate used to assemble to pass laws and try
cases ? " demanded Raed.

" Yas, sirs."

" Well, I declare ! But it is a wonderful place, a terrible place ! Here's where the grim old lawgivers used to
come in the days when Iceland was a free republic, —
eight hundred years ago. Iceland was one of the lights
of the world then. Time has been when these old rocks
have witnessed some of the sternest scenes of history.
Yet where, now, are the multitudes that used to throng
this hoary chasm ? Gone back to the dust; and as for
their souls, I wonder if these lava-ledges are any the
warmer for them, or the wiser for having echoed to their
stern decrees ? "

Nobody being able to answer Raed's apostrophe in
any thing like a philosophical manner, we turned our
attention to the truly remarkable scenery which the rift

4

presents. For the first two or three hundred yards the trail descends, till it strikes the bed of the rift, or *jau* as the Icelanders call it, which is an immense chasm in a lava-field several miles in length. The lava must have been already cooled, or nearly so, at the time the rent occurred. Nothing less than an earthquake could have torn open this long crack. On the left side of the chasm the lava-wall rises perpendicularly for seventy-five or a hundred feet, presenting an inky-black precipice. The right side is not so high, nor yet so regular, though forming a clearly-defined parallel wall. The bed of the crack was thirty and thirty-three paces wide, differing not much from this measurement for more than a mile.

"This looks decidedly volcanic," remarked Kit. "Nothing but fire could have left such marks as these."

"But what split this awful chasm in the cooled lava?" said Wash. "It seems impossible that even an earthquake could have given so mighty a throe."

"Judging from present appearances, this country has been in the way of getting mighty throes, — hot and heavy ones," observed Kit. "Something must have split when all this vast lava-field through which this *jau* runs was thrown out of the earth."

"It is thought that this lava-bed was vomited from the crater of the Skjaldbreid Jokul yonder," Raed remarked; "but it seems hardly possible that so vast a quantity could have run from one crater. Why, here are billions of tons piled up here a hundred feet high. It is more likely to have gushed up through some rent in the earth under us. All Iceland is but a vast hollow

THE ALMANSAJAN.

crater, or a collection of craters, thrust up out of the ocean."

"Somewhere down in the earth beneath, there must be enormous empty caverns where this lava came out," Wash reflected. "Just think of it!—if all this island has risen like a bubble, in the crust of the earth, must there not be a corresponding cavity down under it?"

"Not necessarily," contended Kit. "If the entire inside, or core, of the earth, is a lake of lava, the outer crust probably sinks down whenever any great amount of lava is thrown up to the surface."

"But I don't believe the inside of the earth is a lake of lava," argued Wash; "I never did: neither does Lyell, nor the best scientists of this century. I think the earth has been made up of tiny bits of shooting-stars, meteors, and aerolites, collected into one grand mass during the long ages of the past: I've thought so ever since we saw that big meteor on Mount Katahdin and had that talk about them. The earth is nothing more nor less than a collection of meteoric matter, drawn together, and likewise held together, by the force of gravitation. This process of collection is still going on. Every second, meteors come whirling down like snow-flakes. The earth is growing at the rate of a hundred thousand pounds every day; and that, too, at a very moderate calculation, as we saw when we figured on it that morning. I keep that one fact always before me. That's the keynote to my philosophy. I don't believe there's a spark of fire down at the centre of the earth."

"How do you account for these volcanoes, then?" Kit inquired.

"These volcanoes," exclaimed Wash, with the confidence of a full-grown professor, — "these volcanoes are so many ulcers on the earth's skin, caused by the sea-water finding its way down through crevices and rifts among the rocks. They are due to the chemical action of marine waters on rocks at the depth of a mile or so beneath the surface. The rendings and convulsions which attend irruptions are due more to steam than any other agent, — the steam of this sea-water."

"Well, I'm free to admit that this is too big a question for me," laughed Kit. "Still I think the earth bears evidence of having been in a hotter fire than sea-water and rocks could have kindled. Some time in the past, there must have been hot times. All our granite mountains show it."

"I have no doubt that there have been times when there were more volcanoes than there are at present," rejoined Wash. "Perhaps, too, they were larger and more powerful; but I shall still hold to my opinion, that the earth is but a collection of meteors, and that whatever heat and fire there is beneath the surface is the result of chemical action of one element, or substance, on another."

As we rode on up the chasm, the walls on either side echoing to our talk, the roar of falling waters began to be heard.

"Is there a cataract ahead, Halgrim?" Raed asked.

But Halgrim was at fault with the word "cataract." He shook his head.

"River? is there a river up there?" Raed modified, pointing along the rift.

"Yas, sirs; the Oxeara: it falls over the *jau*."

A few hundred yards farther on we came into view of it, — a strong, bold torrent, sweeping over the main wall down into the chasm at a single plunge with a loud roar. The volume was sufficient to make the earth tremble perceptibly; and a slight mist curled up. For fifty rods below the fall, the stream runs in a continuous rapid adown the bed of the rift; then, leaping through a gap in the opposite wall, passes out into a large lake, of which we had had glimpses several times during the last two hours. Raed pointed toward it, and turned inquiringly to the guide.

"*Vatn*," said he. "*Thing-valla-vatn*."

"This word *vatn* means 'lake,' I believe," Kit remarked.

"And *valla* means 'valley,'" said Wash. "I know so much Icelandic."

"Well, *thing*, or *althing*, was what they used to call their parliament," finished Raed. "So here we have it all translated, — the *parliament valley lake*."

"*Thing* is a rather droll name for a legislative body, rather indefinite, according to our ideas of the word," observed Wash. "*Althing* means 'every thing,' I suppose: queerer still. In like manner, I presume, *almannajau* means 'all men's chasm,' or the chasm where everybody assembled."

"These old Norse words are quite like many of our own familiar words," Raed remarked. "Nearly all of our best, shortest, and most common words are plainly descended to us from the mouths of the Northmen; though our long words are from the Latin."

"That one fact shows the onesidedness of our method of education," exclaimed Kit. "They keep us studying Latin all the best part of our school-days; but who ever heard of a boy being set to study these Northern languages, which are the source and fountain of all the best and most sensible part of our language?"

On coming up where the Oxeara falls into the *jau*, we had halted.

"How much farther are we going to-day?" Wash asked "It's six o'clock now. I want some supper. Let's camp here. There's water, and I noticed grass all along the bed of the *rift*. We can turn the horses out to graze."

"Camp here, Halgrim?" Raed demanded.

"We to the pastor's house can go," said the guide. "It is not a mile. He will be glad us to see."

"Query, does he mean an Icelandic mile?" Kit said.

"No matter," replied Raed. "We don't want to go to any pastor's house. I had far rather sleep under our tent than in one of those wretched *byres* such as we have seen by the way; and I dare say the pastor's house will not be much better. We have got bread and cheese and pressed meat enough; and, if we can contrive to make coffee, we are all right."

"Haven't seen a stick of wood big enough to burn to-day," Kit said. "Don't believe we can get a fire."

"There's a shrub-birch about as big as a high-bush cranberry," Wash remarked, pointing down the bank of the torrent; "and there's some twigs washed high up on the rocks."

Raed jumped off his pony.

"As well here as anywhere," he said. "Going to camp, Halgrim. Take off the pack-saddles. Turn the ponies out. Weymouth, get out the tent: I will help you pitch it."

While they were doing this, Wash and I went off along the rocks to pick up drift-twigs for fuel.

Kit got the hatchet and cut down the shrub-birch, and clipped it into fagots. Altogether we succeeded in gathering up perhaps a bushel of twigs, none of them over an inch in diameter. Halgrim brought a handful of dry grass. Matches were struck; and, with the aid of a bit of old newspaper, a fire was kindled between two fragments of lava. The coffee-pot was then filled from the Oxeara, and, with a generous dip of the fragrant brown bean of Java, was set over the fire on the stones.

It was not without a great deal of patient coaxing that we made it boil. This want of any thing like adequate fuel is a serious drawback to travellers in Iceland. We saw scarcely a shrub over six feet high. The largest tree on the island is an ash at Akureyri, twenty-six feet high. Formerly there are said to have been extensive fir-forests. These have been gradually used up for fuel; and, the severity of the climate increasing each century, they have now ceased to grow altogether. We saw scarcely any thing save low birch shrubbery.

Sitting on our saddles and pack-boxes, each with his pint dipper of sweet, black coffee in one hand, and a hard-biscuit in the other, our party presented a highly-picturesque appearance, no doubt, — the white tent in startling contrast with the black lava-blocks about it;

the foaming cataract, with its sullen plunge, and steady, solemn roar; and, over all, the shimmering, never-setting sun streaming across the awful parapets of the *jau*. Sugar and bread, cheese and bread, meat and bread: we had all the variations which the rather limited number of our provisions would admit of; and, after all, another dipper of coffee. If coffee is really injurious, as many persons contend, I marvel we did not die of it during those weeks in Iceland. We used to drink from a pint to a quart apiece of it twice a day; strong, too, as we could get it from a dipperful of the ground coffee per mess.

Supper over, we sat talking for nearly an hour, — talking of this strange land, and the impressions it gave us. Before us rose the Skjaldbreid Jokul, its summit dazzling white in the sunlight, its black base girt about with dead lavas. Not a tree grows on its gloomy flanks. The green of forests nowhere relieves the landscape blackness. It is this absence of forests which gives such constant impressions of desolation.

"Weymouth, what do you think of Iceland?" Kit demanded.

"Well, sir, it's about the smuttiest, iciest, sootiest-looking place I ever dreamed of. Looks as if it had been burnt over one year, and froze over the next, right along ever since it was" — here Weymouth hesitated for want of the right term.

"Ever since it was *what?*" queried Kit.

"Why — hatched," finished Weymouth in some confusion.

"Oh! that's not the way islands are produced, Weymouth," laughed Raed. "You don't mean *hatched*. Why, don't you know, Weymouth, that the earth is nothing but a big yeast-pot? Islands rise like doughy bubbles on its surface."

Weymouth looked a little puzzled, and was clearly not without suspicion that he was being imposed on.

Shortly after, we dispersed to pick up more twigs for our morning fire. We did this over night in order to have them to *lie on* while we slept. Each of us got a small armful. Wash and Raed found a couple more birch-shrubs. We all gathered what dry grass we could find. Carrying it into our tent, we spread our rubber-blankets over it, and, with our saddles for pillows, lay down to sleep. It was not very cold; and yet, though the sun shone brightly, it was far from warm. With our wool blankets wrapped snugly about our thick woollen suits, we were just comfortable, six in the tent.

Once, ere we went to sleep, we all distinctly felt and heard a deep rumbling sound like the slight shock of an earthquake. I dare say there was nothing very remarkable in this circumstance, when we reflect that the restless Hecla is not more than twenty-five miles distant from the *jau*, and that, within a radius of thirty miles, not less than a dozen craters open down to the fire-caldrons which rage at no great depth beneath. Indeed, the general feeling in Iceland is one of insecurity as to matters underground. I don't know how native Icelanders regard the situation; but a kind of expectant, powder-millish feeling clung to us constantly.

Along in the night, — I call it night, though the sun shone all through it, — Kit waked me. He was sitting up, looking about the tent.

"Where, for pity sake, is Wash?" he whispered.

"Wash?" said I. "Isn't he here?"

"No: he's gone. I believe he went out a long time ago too. Something roused me; then I went to sleep again. He's gone off sure," looking round to his empty berth.

"You don't suppose he's gone on one of his *sleep-walks?*" I said, Wash's noctambulistic infirmity suddenly occurring to me.

"I'm afraid of it!" Kit exclaimed.

We both got up without disturbing the rest, and, pushing back the flap of the tent, stepped out. The sun was behind the lofty wall of the *jau:* its heavy shadow now fell along the bed of the rift. We looked hastily in all directions. There were the horses, — some of them still feeding, others lying down. Wash was nowhere in sight.

"Bet my pony he's gone on a regular tramp!" Kit muttered. "If he doesn't break his neck over these lava ledges, or get drowned in the river, I shall be thankful for it."

We ran out to where the stream pours over the cliff; then, mounting amid the sharp rocks, followed down the rapid. It was a terrible place for boots: the lavas present every sort of cutting, tearing edge, point, and angle. But, working along with great care and hazard, 've came finally to where the torrent whisks out of the *jau* through the lower wall. The noise of the mad cataract drowned our voices, and showered us with spray and damp. It seemed impossible that he could have gone any farther in this direction; for, at the point

THERE WAS WASH SCRAMBLING DOWN.

where the stream rushes through the wall, the crags rise almost perpendicularly for twenty or thirty feet on both sides.

"I guess he went back along the *jau!*" shouted Kit. "I don't believe he came this way."

We were turning back, when the rattling of a bit of lava down the rocks made us pause suddenly. There was Wash scrambling down the side of the *jau* above us, — a side too steep, we thought, to be scaled; and *he had the rifle in his hands too.*

"Heavens!" Kit exclaimed under his breath. "He'll tumble! But don't speak; don't say a word. If we wake him, he will surely break his head."

We crouched down among the bowlders, and watched him. He was jabbering and muttering to himself, — something about a *troll.* (He had been asking Halgrim about *trolls.*) "That sly, wicked *troll!* — that green-eyed *troll!*" He'd "have him;" he'd "*bore him with a slug!*"

Down he came, scrubbing and scrambling over the rough lavas, clinging and holding on like a ferret.

"You don't suppose he would fire at us, do you?" Kit whispered.

There was no knowing. He might take us for *trolls* as likely as not. I advised to keep quiet. He would have to pass us if he kept on toward the camp. We could lay hold of him when he came near us, and take the gun away before he would have time to shoot. He was not long getting down. Such a hurry as he seemed to be in would have been quite amusing, had we not felt so anxious for his safety. Gaining the bottom,

near where we were, he started to run, looking neither
to right nor left. In an instant we jumped up from
behind our rock, and had the somnambulist by the collar.
He was still scolding, and stared at us with unwinking,
expressionless eyes.

"Wash, Wash!" Kit cried, shaking him violently.

He suddenly waked, letting the rifle drop, and lurch-
ing heavily against me.

"What — what — is — the matter?" he ejaculated,
rubbing his brows.

"What's up, anyhow?" getting a little more waked
up.

"Well, that's what we want to know," said Kit.
"What in the world are you doing out here with a gun
in your hands? — cocked too, as I'm a sinner!" picking
it up.

"Where are we?" was the only reply Wash could
make; and so confused and tired out was he, that it was
with no little difficulty that we got him back over the
ledges and bowlders to where the tent had been pitched.
He had not the slightest idea where or how far he had
been, or what he had gone after. We told him some
things he had said about the *troll;* but he knew nothing
of it.

Raed and Weymouth were awake, wondering what
had become of us. Wash had clearly been wading into
the river; for his pants were wet clean to his body, and
his boots were full of water. At no place along the *jau*
were the banks of the Oxara otherwise than ledgy and
precipitous. This fact, as also the dangerous position
he was in when we first espied him, led us to conclude

that he had had a rather adventuresome time of it. A single misstep, either in the water or on the lava-crag, and he must have been swept to destruction.

"Look here, old fellow!" Kit exclaimed a little grimly when we had spoken of these perils. "If this is the way you are going to carry on nights, we'll just tie one of the guy-ropes round your ankle hereafter, my boy: we'll have you anchored before you go to sleep. I don't see how you ever got down into the rapids like that without getting drowned; for the life of me, I don't!"

"Old Scratch always takes care of his own children," Wash laughed, looking a little foolish as the true character of his exploit became apparent to him.

"Well, you won't go *troll*-hunting again without my knowing it," repeated Kit. "I'll have a halter on you, my boy."

I may add that Kit was as good as his word in this. Every night after that, while we were camping out in Iceland, Kit would tie either one of the ropes, or the rein of his bridle, into the strap of Wash's boot, and fasten the other end either to his own boot-strap, or to his saddle when using it for a pillow.

Halgrim had not waked; but as it was now past three o'clock, with the sun high in the north-east, we decided to get breakfast and go on. A fire was built, and coffee prepared, the same as on the preceding evening. The young Icelander was considerably surprised when we called him to breakfast.

"You must ought to wake up me to build the fire, sirs," he remonstrated.

On saddling the horses, two were missing; and Halgrim had to go back nearly to the entrance of the rift — fully a mile — before he overtook them. This hindered us somewhat; and it was not till after seven that we started onward.

CHAPTER IV.

PASSING out of the Almannajau, we crossed the river, and soon came to the pastor's house, or rather collection of houses joined together endwise, sidewise, anywise. There were no less than seven hovels, all connected. I saw but one glass window about the whole establishment. The huts were built mainly of rough stones, with the usual turf roofs of the country. Some of the gables were of boards. It would seem hardly possible that a man of ordinary stature could stand upright in one of these diminutive dwellings. The pastor himself, so Halgrim whispered to us, was standing in one of his numerous doorways, and bowed kindly and gravely to us as we rattled past. He was a small, spare man, with a timid face, rendered sadly sheepish by a close black skull-cap drawn on his head. His long-skirted coat switched his heels. I have no doubt he can talk irreproachable Latin, and is, withal, a very highly-valued

63

and useful man — at Thing-valla. But what Iceland
wants is the railroad and the telegraph, not Latin.

Quite near the pastor's house, and on the other side of
the road, which is here fenced with stone walls, stands
the church of Thing-valla. In New England I should
have taken it for the schoolhouse. It is a quaint little
edifice, about twenty feet square, for a guess, built of
boards, and painted black. The roof, too, is black, being
of boards covered with tarred cloth. The curious little
steeple much resembled a martin-house. Taken together,
the church and parsonage is perhaps the prettiest, neat-
est station in Iceland. The lake, the cataract, the black
wall of the *jau*, and, a little farther on, the rock of the
Lögberg, give it a picturesque as well as an historic interest.

Half a mile beyond the church, we pulled up to take a
look at the Lögberg, or Rock of Laws. Leaving the
ponies in Weymouth's care, we went off across the ledges
to the right. A walk of a few minutes took us down
to the famous spot. No tourist should neglect the Lögberg. It is an interesting locality, remarkable as much
for its craggy scenery as for its ancient story. Briefly,
it is a vast rock, isolated by chasms on all sides, save a
fathom-breadth neck which connects it with the rest of
the lava plateau, in the midst and on a level with which
it stands. The environing chasms are from thirty-five to
forty feet in breadth, and fully seventy in depth. They
are half filled with water; but so limpid and clear are
its depths, that the bottom reflects the light distinctly
through six fathoms. It was into this gulf that the condemned criminal was plunged, with a stone about his
neck.

"I declare! ' Kit exclaimed, shrinking back from the edge of the dizzy abyss, "I should rather be *hung*, I do think, than pitched down there."

"Any thing but hanging for me," said Raed, who is violently opposed to this sort of capital punishment.

"But what an awful sensation it must have given the poor wretch to go whirling down! What a shriek must have risen when he splashed into the water! How he must have struggled with the strangling stone! They say, that, formerly, the bottom of the pool was covered with the white bones of the victims."

"It must have been a very impressive mode of punishment," Raed remarked: "far more so, I judge, than our gallows-scenes."

The Lögberg itself is about four hundred and sixty feet long by seventy in breadth. The flat top is covered with turf. Here it was that the court and judges of the ancient republic used to assemble yearly to try offenders, —murderers, witches, &c. Once condemned, the road to death was a short one. The executioner stood ready to take them from the hands of their judges. The multitudes on the other side of the gulf could but look on in horror. Eight centuries have passed. No one would now mistrust what scenes these old rocks have witnessed; and history but faintly echoes them in legends and mouldy *sagas*.

Kit and Wash made sketches of both the Lögberg and the Almannajau, which, with the cataract, are plainly visible from the point where we stood.

Remounting our nags, we continued on, and in the course of half an hour had entered a tract so wildly deso-

late and black-hued, that we actually shuddered at its
grimness and desolation. Over the whole region a seething lava-flood had poured, and tossed itself in volcanic
wrath. Fury and molten fire seemed to have revelled
and run riot, till the chill blasts of an arctic winter had
turned it to silent stone. Here a black wave, split and
foaming, had cooled as it broke; there a vast bubble of
pent-up gases had burst as it swelled out, and cooled to
glass. Mighty rents gaped beside the path, showing
black depths bristling with jagged spikes and vitreous
points. It took but a small stretch of the fancy to imagine them so many traps set by malevolent *trolls* for
our ruin. A single misstep or stumble of our horses
would have hurled us to destruction. The whole effect
of the scene is suggestive of fiendish malice, — a death-
struggle of the elements. In cooling, the lava had in
many places taken the form of huge frogs, serpents,
shells, and whirlpools. Often we fancied we discerned
the grim faces of gnomes and ghouls staring evilly at us
from out the sooty rifts.

This black and horribly-distorted tract was bounded
on both sides by red, hilly ridges, fenced in by white
peaks shooting up into a dim, leaden horizon.

An hour or an hour and a half took us to the Hrafnajau, or Raven Chasm, which at first seems to present
an impassable barrier. Picture a rent in the lava-field a
hundred and twenty-five feet deep, jagged and chaotic
beyond description, then add a narrow, crooked cause-
way, or bridge, spanning this abyss, all of the wildest,
blackest rock, and you have the Hrafnajau. It lacks the
regularity of the Almannajau; but it is more terrific to
the passer.

Six or seven miles farther on we came to a towering rock, or needle, which Halgrim pointed out with his whip as we rode past an overhanging crag.

"What is that?" said Wash.

"The Tintron, sirs," was the reply. "It has hole in top."

We spent nearly an hour climbing about it. Like all the rocks here, it is vitreous, and of volcanic origin. It rests tilted up on a mound of scoria, a little below the trail. Its height is about thirty feet, but so jagged and notched, that it can be climbed, if one has the nerve to attempt such a feat. The most singular feature about it, however, is that to which Halgrim had referred, — the hole in the top. This hole is about three feet in diameter, and seems to have been a miniature crater. Kit threw several pebbles into it. They seemed, he said, to go rattling down to an unlimited depth. For several seconds he could hear the sounds till they died away in the black depths. The Tintron gave us the impression of having been a diminutive volcano.

Going back to the trail, we took a lunch, and, after a short rest, went on descending the side of a ridge into a valley where there was a small, rapid river, which we forded, drawing up our legs like crickets to keep them dry.

Iceland abounds in rivers. For the size of the island, and the extent of country they drain, they are very large and rapid. In nearly every case we were obliged to ford them. Iceland hasn't got to bridges yet. Hold! I forgot the Bruara, which we crossed during the second afternoon of our trip to the geysers. This stream has a bridge of a rather novel sort. On coming to the bank, we supposed it was a ford, — a very dangerous one it looked too. The current toward the middle ran like a mill-race, foaming and roaring among rocks. The trail led directly down to the water, however; and Halgrim drove in without a word.

"Is it really safe?" Raed shouted.

"The Bruara, sirs," replied the Icelander, his voice half drowned by the torrent.

There wasn't much consolation in these words for us, who judged the Bruara by its looks. Raed, who was a little ahead of the rest of us, prudently halted to see how Halgrim got through. Somewhat to our surprise, as the leading pony approached the middle, we heard the clattering of his hoofs on planks, and saw them all walking quietly over, with just their fetlocks buried in the water and foam.

"Some sort of a bridge there," muttered Raed, spurring in.

The rest of us followed. Sure enough, out at the channel, where the stream ran madly between two ledges, there was a plank platform spanning the deeper part, and pinned securely to the rocks on both sides. At either end of it the water poured past up to the ponies' bellies, and ran in a streak of foam over it. Perhaps the stream was higher than usual; but it struck us as a rather strange bridge.

"Do you call that a bridge, Halgrim?" Kit asked when we were safely on the other side.

"The Bruara — the bridge," said he.

"Does Bruara mean bridge?"

"*Yawh.*"

Halgrim said *yawh* for *yas* occasionally.

We passed but three huts during the afternoon, and saw but one person, a woman, we thought, driving sheep at a distance. To the right the country sloped off to a wide valley, with here and there a thicket of low brushwood. To the left rose a long volcanic ridge. Beds of moss, white as snow, lay interspersed amid the ledges.

About five o'clock, Raed pointed suggestively to a column of white vapor rising over the rolling ground, seemingly six or seven miles away.

"Is that the geyser steam?" I exclaimed.

"I expect so," said Raed.

We all gazed toward it with queer feelings struggling up. Could it be true that we were really so near the famous geyser of which we had read and heard so much ever since we were mere children? — those wonderful geysers, a cut of which had adorned the first pages of our earliest school-geography? An hour more, and we should be there and see the real thing. No picture this time, but the genuine hot-water fountain! Why, it seemed almost too good to be real.

Halgrim saw us looking off, and immediately said, "The geysir, sirs."

I think he had meant to get us up nearer before pointing them out. Rude as are the most of these Icelanders, they have a sort of national pride in their wondrous hot springs, and expect travellers to be greatly astonished at them.

"They are not to be much seen so far," he turned to explain after we had ridden a few rods in silence.

"Do they shoot up big, Halgrim?" Kit asked.

"They shoot up very high!" he exclaimed enthusias-
tically.

"How often does the big one go up?" Wash asked.

"Once in three and four days now, sirs."

"So long as that?"

"Yas: sometimes *he* is a week now. He used to go
up often, — go up every six hours: now he is slow. He
does not so high go up, either; not so high; not so hot;
not so much."

"That agrees with what we've read," remarked Raed,
— "that the Great Geyser is dying slowly. Some change
is going on in the funnel, or pipe. In time, these irruptions will cease altogether."

"How long has the Great Geyser been running?"
Wash asked. "Does anybody know?"

"It began in the year fourteen hundred and *something*," said Kit. "That's all I can tell relative to its
birth."

"I never could learn that any one knew just the year
it first burst out," Raed observed. "It was some time in
the fifteenth century, as Kit says."

Wash remarked that he had read that two Danish
travellers, named Olsen and Paulsen, saw an irruption
of the Great Geyser, during which the water was thrown
three hundred and sixty feet high.

"Guess they stretched it a little," was Kit's comment
on this story.

"How high does the water shoot up, Halgrim?" I
asked.

"Goes up seventy feet," was the answer.

" I wonder if he means Icelandic feet," Wash queried.
" My old geography said it went up eighty and ninety
feet."

" Bet you we don't see it over sixty !" Kit offered.
" But it will be a joke if we should have to wait a week
for it."

" I mean to see it, now I've come so far, if I have to
stay here a fortnight," said Raed.

" Of course."

" Do travellers ever have to go home without seeing it,
Halgrim ? " I asked.

" Sometimes, sirs. The Prince Napoleon came to see
it. He could only stop two days. The geyser would not
go up. He had to go back without seeing it."

" That was rough on the prince !" laughed Kit. " The
geyser ought to have been more polite."

" Well, the prince has met with more serious ill luck
than that since then," Wash remarked.

" How many geysers are there, Halgrim ? " Raed
asked.

" There is the Little Geysir, the Great Geysir, and
the Strokhr (pronounced Strokker). Then there be
others many, smaller."

" The word *strokhr* means 'churn' in Icelandic," Kit
remarked. " I saw it so stated in Mr. Metcalf's works.
The Icelanders called it a churn, because they fancied
that the motion of its waters resembled *churning*."

" Strokhr is the one you can make go up by throwing
in sods, isn't it ? " Raed inquired. " Do they throw
sods into the Strokhr, Halgrim ? "

" Yas, sirs : that will *him* make throw up."

"Well, Halgrim, set those forward ponies galloping," exclaimed Kit. "Let's be getting on. What did you say was the name of that long mountain off there ahead, — that jokul over there, Halgrim?"

"That is the Langarfjal."

Riding on over low hills and slight green valleys for three or four miles farther, we came to another bog lying along the base of the Langarfjal bluff. Following along this bog for some little distance, we crossed it without difficulty at a favorable spot near the huts of a shepherd family. Several of the children stared after us from the doorway. Such great round faces and tow heads are not to be seen out of Iceland.

Spurring on, we soon drew near the slender column of steam which marks the position of the Little Geyser. Eager and expectant enough we were, no doubt. Up, up, up, streamed the steam-cloud, and a rumbling, restless sound rose from the ground. Raed was a little ahead; and, on coming within half a dozen rods of the steam-jet, he jumped off his pony, and ran for it: the rest followed close behind him. The earth was red here, and gave back hollow echoes beneath our footsteps. We all seemed to feel the near, alarming presence of some terrible agent, the outbreaking of which is always attended by peril to man. Beneath our very feet throbbed and raged those olden fires which have burned and fused the earth as in a crucible.

And yet the Little Geyser has nothing very impressive in its performances otherwise than these feelings which it gave us. It is a little round hole in the earth, two feet in diameter, and, as we afterwards found, about thirteen feet down to where it crooks off laterally. The steam gushes up blithely; while a hollow, boiling sound tells of the volcanic fire-heat below. There was no water in sight in the shaft on the afternoon of our arrival. We could not see down more than five feet for the steam; but on several occasions during our stay the boiling water gushed up, though never to the height of more than three or four feet, overflowing, and running down the side of the slope. Quite near the funnel of the Little Geyser there was a puddle of mud in violent agitation. It was thick, black, and viscous as treacle, and kept swelling up in bell-shaped bubbles that occasionally burst with a dull *thut*, letting out tiny whiffs of steam. From the Little Geyser, the slope on which this wonderful system of springs is situated rises gradually to the steep side of the Langarfjal ridge. The whole number of the springs is about forty, little and great. The slope, or, as some travellers call it, the plateau, out of which they gush, is about four hundred yards long, and of varying width. Whatever the original soil or rock may have been, it is now strongly saturated and charged with mineral matters brought up with the water. The yellow hue bespeaks the presence of sulphur; and a great deal of what seemed to me burnt clay and crumbling trap-rock lay about. Evidently at some former period the whole slope has been subjected to an intense heat. It seemed to have been baked to a cinder. Steam and smoke, accompanied by odors of sulphur, steal up from numerous little fissures all about.

After eying Little Geyser for a few moments, we

turned all together, and ran for the Strokhr. Some
competitive, boyish spirit made us each anxious to get
the first sight of each spring, or at least not let the
others be ahead. Arrived at the mouth of the famous
churn considerably out of breath. An angry growl, followed by a swash and hiss, led us to halt rather hurriedly
several yards from the orifice.

"Going to erupt?" Wash queried.

The ominous sounds ceased directly.

"But isn't it a big tube?" Kit exclaimed, stealing
cautiously up.

The mouth of Strokhr is not raised like that of the
Great Geyser. It is simply an opening in the siliceous
rock, which some have compared to a saucer, though we
failed to see the resemblance. It is not quite circular.
We found its average diameter to be nearly eight feet
at the surface, diminishing as it descends, till, at a depth
of thirty feet, it is scarcely over two feet across. The
total depth of the pipe (as far as our stone and line
informed us) is between forty-four and forty-five
feet.

When we first looked into it, the water was boiling
and swirling at a depth which we *guessed* to be nineteen or twenty feet. Steam whirled up spitefully at
times, though never very densely.

"Considered as a place to tumble into," Kit remarked, eying the footing about the brink, "it hasn't
an inviting look."

"You are right," heartily responded Raed, drawing
back, as if not quite at ease about Wash, who was peering
over his shoulder. "One souse into that would boil a

fellow like a lobster! Never get out alive. Ought to be a railing round it — for somnambulists."

" I'll see that Wash is picketed all secure to-night," Kit said.

" Now for the great one!" shouted Wash, quite indifferent to these saving resolutions of his friends.

We all heeled it again off diagonally up the slope to the north-east, where a thin, steaming column, about which the hot air quivered, denoted the locality of the world-wide wonder. A scamper of a hundred and fifty yards took us to the foot of the mound of rock, in the top of which is the mouth of the geyser-tube. The mound itself is raised about forty feet above the surrounding slope. We climbed up, — it is not very steep, — and found ourselves standing on the rim of a huge circular basin of brown rock, as smooth inside as if sandpapered every day. This shoal basin is fifty-two feet in diameter from north to south, and sixty from east to west; not quite circular, yet so nearly so as to have that appearance. The mouth of the geyser-tube is nearly at the centre of the basin. At the point where the tube opens into the basin, it is ten feet in diameter, and, so far as we could see, an almost perfect circle. The depth of the tube is seventy-five feet. Letting the sounding-line drop in on one side, it is seventy-four feet, and on the other side seventy-six feet: I therefore give the average depth at seventy-five feet. The tube does not contract like that of the Strokhr, but maintains an almost uniform diameter as far down as can be seen. The tube, like the basin, is of siliceous stone, polished, and smooth as glass. As a work of art it would be wonderful; but as a work of nature it is indeed so.

When we first climbed the mound, the basin was quite dry; and, after surveying it in a kind of first-view wonder for some minutes, we got down into it, and walked up to the mouth of the pipe. It was not without a feeling of awe that we ventured to gaze for the first time into this mighty *boiler*. I saw my own feelings reflected in the faces of my companions as we trod the hard echoing floor of the basin, and cast our eyes timidly into the huge, gently-steaming orifice. It was nearly full, and seemed quite limpid and clear. We could see far down the sides. It did not boil like the Strokhr. It seemed in a very quiet and contented frame of mind; though its surroundings were rather suggestive of a violent and explosive temper. In fact, the general aspect of all the springs that afternoon was peaceful. Some queer noises were from minute to minute audible underneath. "The old shebang hadn't stopped," so Weymouth thought. Business had not been suspended yet. The *trolls*, fire-fiends, or whoever run the *establishment* there in the shadow of the Langarfjal, were clearly at home, and had steam up; but they kept their doors closed. Kit thought that they might be "taking account of stock," "greasing the big wheel," or something of that sort.

Well, it would be no use trying to hurry the Great Geyser, who, according to all accounts, was a very disappointing old chap, always snubbing and slighting tourists, including his Excellency the Prince Napoleon But we could wait, and had started with the intention of waiting till the old fellow would be glad to heave up or " bust."

"But we can keep Strokhr to play for us," Wash reflected; "unless, indeed, he has lost his antipathy to soda."

CHAPTER V.

GETTING out of the basin, and going down the mound, we came, at a distance of about fifty or sixty yards from the Great Geyser, to the little grassy spot where tourists usually have encamped. More than a dozen small, partially-filled trenches showed where their tents had been pitched. We called to Halgrim to drive up the pack-ponies : which he did with some trouble; for they heard the ominous grumblings from beneath the plateau, and kept snorting and halting as the guide urged them up. Under their hoof-falls the ground resounded hollow and cavernous. They were afraid. Perhaps it was hardly safe to drive them over it.

Throwing off the pack-boxes and saddles, we had Weymouth and Halgrim (at the latter's suggestion) take all the horses back to the shepherd's house to see whether he would pasture them while we were at the

78

geysers. We also directed them to borrow a spade, buy milk and whatever other clean-looking eatables the people might have and be willing to part with. There was an old spade, broken and bent up, lying near the mouth of the Strokhr. With this, while they were gone, we cleared out one of the tent-trenches, and pitched our own tent on the site of some other party's labors, — the Prince Napoleon, for aught we knew to the contrary.

It was now about seven o'clock. We had taken nothing save a dry lunch since morning, at five o'clock.

" No fuel here," Wash remarked rather dubiously.

Raed laughed.

" Wash has forgotten the grand culinary advantages which the geysers offer to their visitors," said he.

" You aren't posted, Wash !" cried Kit. " Don't you see it ? Hot water is the great staple here ! "

" Let's have coffee instanter !" exclaimed Raed. " All we have to do is to let our coffee-pot down into Strokhr or the Great Geyser: it will boil there in a jiffy."

" There's quite a steam rising up there at the foot of the lava-bluff," Kit said, pointing to it. " I'll go reconnoitre. Perhaps we may find the water boiling in a less dangerous place than the Strokhr. It would take all our guys to lower our coffee-pot down into *that*."

He went off, and a moment later called to us. Raed and I went up to where he was standing on the brink of a large hole, twenty-five or thirty feet in diameter, filled to the very brim with steaming water.

" Here's a big boiler for us," Kit observed.

" Is it hot enough to boil things ? " I asked.

"Try it with your finger," said Kit.

Stooping, I thrust in my fore-finger, and had it uncom·
fortably scalded, though I jerked it out instantly.

"This is the place to do our cooking," remarked Raed.

"There's another pool quite like this just beyond,"
sai↲ Kit, pointing a little farther along the foot of ꜩe
crag. "They seem to be huge bowls in the solid rock.
Can't be less than twenty feet deep ; but you can see the
bottom, for all that. Don't you see that old tin kettle
lying on the bottom down there ? Somebody lost it in
there. If we had a hook, we might fish it out. And,
out in the middle there, don't you see a dark spot on the
bottom ? I think that is the mouth of a tube, like the
Great-Geyser tube, running down nobody knows how far.
These pools are the basins to the tubes."

"Looks like that!" Raed exclaimed. "These pools
were geysers once."

"I think so," said Kit.

"Are not these large still pools of hot and warm water
what the Icelanders call *laugs ?* " said I.

"Well, not so hot as this," said Raed. "I think that
they call a fountain a *laug* when it is warm and just
right to bathe in ; but, when they are scalding hot like
this, they term them *hvers.* In the same way, when a
fountain spouts up hot water, it is a *geyser ;* but, when
it merely sends up clouds of steam, it is a *reykir ;* and
a mud pool like that down by the Little Geyser, that
is a *náma.* I think that is what I have heard. We
will ask Halgrim about it."

Kit had gone down after the coffee-pot and the coffee.
Wash came back with him. They dipped in some of the

boiling water, and took turns holding it down in the pool till it boiled, or scalded, and steeped.

Raed and I started to go back to the pack-boxes to get the bread and sugar ready, when we espied Halgrim, Weymouth, and two other persons, coming. The latter were carrying great bundles of *something*, the exact character of which was not at first apparent. Halgrim had a tin pail of milk, and a spade: Weymouth had a smaller dish of cream, and a package wrapped in cloth.

"It's all right about the ponies," cried our sailor companion. "Old fellow here offered to take 'em all a week for five dollars, — fifty-cent dollars. Dog-cheap, I said. And we've got him and his cub of a boy to bring us up some *hay* to *lay* on nights."

"*Hay!*" Raed exclaimed. "I should call that stuff brushwood."

"Well, so should I," replied Weymouth; "but *he* calls it *hay*. It is what he feeds his cows on, anyway, in the winter, and horses too. I thought it would do to lay on, — better than the bare rock: it's pretty *springy* stuff. We got those big bundles for another fifty-cent dollar."

The old Icelander and his son kept close to Halgrim, and, at a word from him, laid down the bundles. Not a twig of it was *finer* than a pipe-stem. They both had fur caps, loose cloth frocks, or jackets, and trousers of sheep-skin. They were plainly honest, kindly folks, who took a friendly interest in us; and though it was impossible to talk with them, yet they asked Halgrim a hundred questions about the "young gentlemen who had come clean from down on the other side of the world."

6

Their faces were broad and good-humored. "Better specimens of humanity than one would expect to meet in a region so desolate and outlandish," I heard Raed saying to Kit.

We had Halgrim give the old shepherd five rix-dollars, and ask him to send out four quarts of milk to us each day while we remained at the geysers. At first, he quite refused to take what seemed to him so large a sum (two dollars and fifty cents) for so small a service; but on Halgrim explaining to him, that, in the great and rich land from which we came, "two dollars fifty" was esteemed very lightly, he took it: and, with many backward glances, the two went off to their humble *byre*.

"That's what I call a rather good thing," Kit said.

"What is?"

"Why, to find a man with a conscience about taking a big price when it comes in his way," Kit replied. "I declare, fellows, this poor old shepherd makes me blush when I think of our people, — the most of them."

Four pack-boxes laid side by side on a patch of level turf, and covered with one of our rubber blankets, made our table. The biscuits, the sugar, the meat, the coffee, the milk, and the cream were set on; to which Weymouth added his package, casting a sly glance around where we were seated on our saddles.

"What have you there?" demanded Kit.

"*Skyer*," was the answer.

"*Skyer!* Well, what's that?"

"Oh! it's a cheesy sort of stuff. They had nothing else to sell, besides their milk: so I bought some. It looks decently clean," unfolding the cloth, and disclos-

ing a yellowish mass of what turned out to be curd, which
takes the place of pressed cheese with the Icelander.

Cream in our coffee was something of a luxury; milk
to drink was another; and biscuit and *skyer* was a nov-
elty, if not a luxury. We supped in all the barbarous
profusion of a camp of savages; and it was not till after
nine that the feast concluded.

The wind had begun to blow in chilly gusts. Dark,
wild clouds flew swiftly over from the north-east. The
crags of the Langarfjal sighed dismally to the fitful
blasts. We tightened the guy-ropes, and, carrying in the
pack-boxes, piled them in a tier along the windward side
of the tent. The *hay* was then spread out on the ground.
It made a bed fully two feet thick all over the interior
of the tent, and, with the rubber blankets spread over
it, promised a very tolerable couch — for tourists.

This done, I proposed to turn in. We had not got
more than three hours' sleep the preceding night; and
what with riding over lava-fields, fording rivers, and
climbing about the Lögberg and the Tintron, the reader
can readily imagine how tired we might be. But no:
nothing would induce Kit to go to sleep till he had seen
Strokhr heave up.

"But, Kit," Raed argued, "we are so tired! and we
can just as well make it spout to-morrow."

No: world might come to an end before to-morrow.
Iceland might blow up; or, at least, this geyser region
might. Shouldn't be surprised if it blew up any minute. 'Twas a queer sort of a place, to make the best of
it, — not to be relied on. He hadn't come all the way
from Boston to run any risks of not seeing the Strokhr
spout.

"But, if the world comes to an end before morning, it won't make any great odds whether you see Strokhr or not," I ventured to suggest.

Yes, it would. Why not? The fun of seeing it would be all the same to-night if the world did collapse by morning. He wasn't going to be so foolish as to lose the present chance by taking a snooze of six or eight hours. Something might happen. Human affairs were very uncertain. He meant to see Strokhr forthwith.

This, and a lot more of similar nonsense, delivered in the gravest possible manner.

The fact is, Kit never gets tired. I never saw just such a boy. Whether he sleeps or not doesn't seem to make any great difference with him. I suspect he has got what my grandfather used to call an "iron constitution;" something of that sort. So, now, while the rest of us were quite tired enough to go to sleep, he was all impatience to see Strokhr perform.

"But there will have to be lots of sods cut," said Wash.

"I'll cut 'em;" and, seizing the spade Halgrim had brought, he set off.

Strokhr was about a hundred yards from our tent. Weymouth secured the old spade, and started after him. Of course, we would not go to bel, and have Strokhr heave up without seeing it. Throwing off our weariness, the rest of us followed; and soon we were all busy as "Paddies," cutting and carrying turf, and piling it up in a great semicircular heap all around one side of the mouth. Strokhr growled some, as if he knew from past experience what was coming; and, on looking down the pipe,

I saw that the water had risen several feet since six o'clock. Halgrim told us that the peasants said that it had been known to go up of its own accord; though no tourists have seen such a phenomenon. We worked for half an hour, I apprehend, cutting up the fine, green turf into sods, and carrying them along, till the heap was two or three feet high, and extended nearly half around the mouth of the tube.

"There's enough," Wash pronounced.

Raed and Halgrim thought there was enough: so did I, for that matter; though, being pretty tired, I didn't much like the labor of tugging it.

"No, no!" shouted Kit, staggering up with his arms full, — "not half enough!"

"They do never put more in at once time," mildly remonstrated Halgrim.

"No, Kit. Here's enough in all conscience!" cried Wash.

"I tell you there's not half what I mean to give the old chap. If three sods make him a little sick, three hundred will make him a good deal sicker. It's just as it is with an emetic: one *boll* of lobelia will make a fellow *squeamish;* but a whole cupful is what makes the tears come. I mean to give him a dose that will make him remember the 'young yachters' to his dying-day."

"But we may choke him up entirely," Raed observed.

"Nonsense! He's got awful muscles to his stomach. Let's make him sicker than ever he was in all his life, — 'sick as a horse.' He'll spew all the higher for it. Come on!"

Thus exhorted, we all fell to work again; and, for

twenty or thirty minutes, nothing was heard save the
nacking and grating of the spades, and occasional un-
easy growls from the subterranean boilers.

"There!" assented Kit at last, when the heap of
clods was near four feet high, and piled all about the
mouth. "Can't be much less than two tons of it.
We've *skinned* two or three square rods of ground, at
any rate. Doubtful if there was ever a bigger dose ad-
ministered at one time. All ready now! Push 'em
over!"

We all pushed at the tottering heaps. Over they
went, and plunged *slap-dash, slap-dash,* into the steaming tube.

"Leg it now!" Weymouth shouted, running off with
the spades.

Halgrim was walking off. We followed him for a distance of thirty or forty yards, and stood momentarily
expecting to see it spout up. A minute passed.

"Does it go up quick, Halgrim, — right off, — soon?"
Raed demanded.

"Yas, sirs; right soon!"

But it did not immediately. Another minute passed
We could hear it growling in deep bass, but not very
loudly. Kit began to edge up. An overpowering curiosity to look in and see what was going on drew us all
cautiously forward. Kit was already peering furtively
into the shaft. The water, discolored and inky from the
black turf, was *whopping* over and over, splashing
against the sides of the *churn* with an ominous hollow
sound. Suddenly there came a louder growl, a rumble,
and a tremendous splashing. A vast escape of steam,
smelling strongly of sulphur, flew up.

"It's coming!" yelled Wash.

"*Ah-r-r-r, ah-r-r-r!*" Halgrim shouted.

We ran precipitately off some twenty yards, and turned — to see that nothing had come of it.

"False alarm!" said Kit.

We began to walk back. Another rumble and loud hissing made us pause. It subsided, however, and we went up to the mouth again: we were growing bolder. Halgrim evidently didn't know much about it.

"I'll not run again till I see it start," Wash muttered. "Time enough to get out the way then, I imagine."

A threatening roar not to be mistaken drowned the latter part of this sentence: a swirling, *sudsey*, rushing noise accompanied it. Before we could even dodge, up went the steamy, roaring, black jet past our very faces! To say that we jumped and scampered would but faintly express the incontinent haste with which we *vamosed*. In an instant we were a dozen feet away; yet, hurried as was our departure, the falling torrent besplashed us. We *heard* it descending. Several of the hot, steamy sods struck down in front of us as we sprang off. One of them hit Raed, smearing him with mud. Where the spatters hit our hands and faces, they felt pretty hot; but they did not scald through our clothes. Our sudden retreat tickled Weymouth very much. He fairly doubled up with laughter, mingling his lusty *haw-haws* with the rush and plunge of the *churn*. Halgrim, on the contrary, saw no joke at all in this circumstance, and stood regarding us with a very grave and concerned face. That is the difference between a Yankee and an Icelander.

Going a safe distance, we turned to see the display
I recollect that my first feeling was one of astonishment
as to its height. It loomed up amazingly. There
seemed to be scores of separate individual jets shot up
so rapidly as to all be in the air at once. Some of these
went far higher than others; and, all shooting up and
falling at once, they were looped and intwined on a vast
and intricate scale, making altogether an immense
column, which at the base could not have been much less
than fifty feet in diameter, gradually tapering to the
head. A rapid, coughing noise accompanied the jets.
The roar and rumble from the earth continued. There
was also a continuous hiss, like that of the escape-pipe
of a large steam-engine. The splashings of the falling
water were heavy and ponderous: the earth trembled
perceptibly beneath them. The vast amount of steam
which would naturally fly off from so mighty a column
of boiling water in a chilly air adds greatly to the effect.
The whole scene to the leeward is filled with white rolling clouds, against which the water-jets show black as
ink. The ground was deluged for twenty yards from the
mouth of the tube. Anon a stray jet would fall far out
toward us. We afterwards found some of the sods fully
seventy feet from the mouth. The most of them, however, fell directly down, and were swept back into the
pipe by the receding flood, with loud gurglings.

It was a grand spectacle, — far grander than this
simple hackneyed assertion will give the reader any idea
of. It was grander, too, than we had expected — from
Strokhr. Taken with the wild, dark clouds, the shadowy
crags of the Langarfjal and the gloom of the evening, --

WE TURNED TO SEE THE DISPLAY.

for it was now past eleven o'clock, — it made a scene which will never fade from my memory. We all quite forgot our weariness in the sombre grandeur of the sight.

The fountain continued to play for six or seven minutes, I judge; though we had none of us thought to time it by the watch. It then subsided quite on a sudden, with a few straggling jets and a loud *sucking* noise, as if the pipe were empty. Nothing was then to be heard save the gurgle of the returning waters pouring back down the tube; for the rock about the mouth is sufficiently saucer-shaped to train a great part of the water back.

"Well, he's done for to-night, I take it," Raed said, drawing a long breath. "It's a great thing, this Strokhr! — worth coming to see. Wonder if the Great Geyser can beat him."

"Wish the great one would take a notion to go up now, so we could sort of compare the two performances," said Wash.

"We might manage to have 'em both go off at once," Kit thought, "by cutting a lot of sods, and piling them ready about the Strokhr's mouth. Then, as soon as the Great Geyser showed symptoms of an irruption, we could pitch in the sods, and so have them both going at once."

"Not a bad plan, Kit!" Raed exclaimed.

"We'll do it," said Wash.

"I don't see why we cannot," Kit remarked.

I thought the only difficulty would arise from the fact that we none of us knew the geyser's symptoms.

Raed replied, that, if we remained to witness a second irruption, we might, by observing the signs which preceded the first, be able to manage it.

We had turned to go back to our tent, when a deep growl from the exhausted Strokhr made us turn.

"Sick again!" laughed Kit: "hasn't got over it yet!"

Another deep rumble began, followed by a prodigious gurgling, seething, and hissing, as a great gush of steam puffed up.

"Will he heave up again, Halgrim?" I asked.

"Yas, sirs: many more times. Heave up worse than at first time."

"Is that so?" exclaimed Kit. "Then it will be all the easier managing to make 'em both heave at once."

A moment later, the same threatening roar made the plateau tremble afresh; and up went the inky flood a second time, blacker and more fearful than before. I have no doubt that some of the jets this time leaped full seventy feet in the air. Seventy feet, too, of perpendicular elevation, is a great height, — greater than I had realized when reading of the geysers during our voyage to Reykjavik. Relative to the heights to which these fountains play, I may remark, that, two days after, we made a rough measurement of the height of the Strokhr jets; using our tent-pole and fish-lines to form a triangle, along the hypotheneuse of which we sighted to the head of the Strokhr column, according to one of the most readily-effected trigonometrical methods. The height of the jets was thus computed at sixty-three feet and seven inches. But we were all agreed that the

irruption on the night of our arrival, when we had in
so many sods, was considerably the highest: Raed estimated it to be ten feet higher.

The second "heave-up" lasted five minutes and a
quarter by the watch, as Wash reported; at the end
of which time the fountain ceased after a few gruff sobs
and a long-drawn growl. The ground all about the
mouth, and far down the slope, was left steaming pro-
fusely.

We went to bed immediately on our *hay* and blankets, and in half an hour were drowsing off, when a third
rumble and roar roused us up.

"It's Strokhr, sicker than ever!" Raed muttered.
"Hark! what a splash and hissing!"

It was impossible to resist the fascination of getting
up to see it again. Tired as we were, we could not yet
lie there contented while a "heave-up" was going on.
Once up, we watched till it subsided; and so another
half-hour was gone before we were fairly abed again.
By the time we were well asleep, a fourth rumble and
roar awoke us. In short, I doubt if any of us, save Halgrim, slept two hours that night: for the irruptions continued, at intervals of half an hour up to an hour, till
after five o'clock in the morning; and, on looking into the
funnel at a little after six, I saw that the water was still
turbid, and swashed uneasily about, showing that the
dose still produced unpleasant effects. Two tons of sods
made a more than ordinary severe emetic, even for a
Strokhr's stomach. Kit said there were nine irruptions
in all during the evening and night. Altogether, we
had all we wanted of Strokhr · it was quite a relief to
have him stop.

About half-past six, it began to rain violently The wind blew hard, and howled down from the dark gorges of the Langarfjal in fierce, wild gusts, making our canvas swell and flap. The gusts of rain, hurled violently against the tent, sent the spray through in a perfect mist. To remedy this defect, which promised us a drenching ere the storm was over, Kit and I took the "rubbers" from off the *hay*, and, putting on our great-coats, ran out and threw them over the top of the tent, making them fast with bits of fish-line through the eyelets. This stopped the spray, warding off the rain. India-rubber and its twin-brother gutta-percha save us many a wet back first and last.

"How about breakfast?" says Wash.

"I move you we have a soup," suggested Kit. "We've macaroni, meat, and biscuit. What more could you want?"

"Soup it is, then," said Raed. "We can flavor it with cheese, and *skyer* to boot. Unpack the tin kettle, Halgrim, and get out the fixings."

"But who will be martyr enough to sit out there in the rain to hold the kettle in the pool while the soup cooks?" I queried. Whoever has that job is sure of a ducking. There's the coffee-pot too: somebody's got to hold that in."

"Well, it may as well be me as any one," said Weymouth with a sigh of resignation.

"And I the coffee-pot will hold," offered Halgrim, with a rueful glance out past the flap of the tent.

Of course the rest of us offered no objection to this proffer: we were quite willing.

The ingredients for the soup were thrown into the kettle. Salt and pepper were added; and Weymouth, with one of our great-coats buttoned about him, and his hat pulled down, was started out with it, having instructions to first fill the kettle half full of the boiling water, and then hold it well down in the pool till it had boiled fifteen minutes. The coffee-pot was then charged, and Halgrim sent out on a similar errand.

'Twas a curious way to get breakfast. For half an hour we could see them crouching patiently on the edge of the *hver* like two big bull-frogs beside a puddle, the rain pouring down like a shower-bath. And it wasn't a warm rain, either: it seemed very much like one of our late October rains at home, when we see ice the next morning.

On coming in, both Weymouth and Halgrim were chilled and shivering; and we in the tent had to knock our toes to keep them from aching. But the hot soup and steaming coffee warmed us up wonderfully. The soup itself was so excellent, that Kit magnanimously offered to *hold in* another kettleful for dinner.

It rained till after five o'clock in the afternoon. We kept as snug as possible; slept a part of the time. Wash and Kit finished up a sketch they had made of Strokhr, to adorn my narrative (which, as I am well aware, is sadly in need of adornment).

The storm cleared off with a smart, chilling wind from the north-east. It was cold, — biting cold. We hung up our overcoats inside the tent, and shook up the *hay*, making as cosey a nest as possible; into which, after supper, we all *crawled*, and lay talking, joking,

and snoozing, — waiting, waiting, waiting for old G. G.
(Great Geyser), as Kit had abbreviated him, to favor us
with a performance. But as yet we hadn't heard a
word, not even so much as a whisper, out of his mouth
Strokhr growled and tossed all day; but G. G. steamed
as noiselessly as a *laug.* The old shepherd had
told Halgrim that he had, if he remembered correctly,
seen the water shooting up two days before we came.
Possibly, too, it had gone up the night before we
came. Halgrim also informed us that there was always
a good deal of disturbance and subterranean commotion
prior to an irruption. Clearly we had no great reason
to look for a performance right off. Raed thought, too,
that so much cold rain-water falling into the basin and
running down the tube might delay it. So we gradually
went off to sleep, and, having once got fairly at it, slept
soundly till after two o'clock, morning, when a noise
altogether different from Strokhr's growls aroused me
on a sudden. Raed was awake, and up on his elbow,
listening.

"Is that G. G. ?" said I.

"I think so," he replied. "Heavier than Strokhr's
voice; comes from up back of us too."

It sounded very much as I should suppose a fifty-
pound cannon would, if fired half a mile *down in the
ground,* — a dull *bum-m-m-m-m!* followed by a slight
jar of the earth. These reports succeeded each other
at intervals of from one to three minutes.

Bum-m-m-m!

Bum-m-m-m!

Bum-m-m-m!

We lay there listening to them for twenty minutes or half an hour.

"That's G. G. fast enough," said I. "We ought to wake Kit: he never'll forgive us if we don't."

"That's so. — Kit! Kit!"

At the sound of his name, that indefatigable young worthy started nervously up, broad awake the first thing.

"G. G.!" said I significantly.

"Is it, though?"

"Listen!" says Raed.

In a few moments the geyser let off another of his minute-guns.

"Yes, sir! That's *him*, sure!" Kit exclaimed, getting up to look out. "Sounds a good way *down*, though, doesn't it?"

Raed said that it now sounded twice as near and heavy as when he had first waked.

Kit went out to reconnoitre. On parting the flap of the tent, the cold air rushed into our warm nest. The wind was blowing sharp and hard from the north-east. It was cloudy too: straggling, dark fogs were driving by, not very high up. The morning light broke in but faintly through the cloud-banks that lay along the east and north-east. As we were peeping out, a heavier, louder *gun* from the geyser-boiler broke alarmingly on our ears; and Kit came back to report that the water was within two feet of the top of the tube, and that it kept heaving and slopping up into the basin.

"It bubbles as smartly as did our 'poison spring' at Katahdin," he added. "Steam-bubbles, I think, — steam and gas; for I smell sulphur just as we did at the Strokhr."

The rest were rousing up. Either this last *boom*, or else our talking, had waked them. We all turned out, and climbed up into the basin to see what took place down the funnel when the explosions occurred. And *in* this instance we had our curiosity immediately gratified. Scarcely had we approached the funnel when another heavier report boomed from below, and the water bulged up to the height of three or four feet above the mouth, and came dashing out into the basin. We scampered off to the rim, which we mounted in great haste, just as the water was gurgling back.

"We had better keep away," was Raed's comment. "It would be no joke to get caught here in the basin when a big jet comes up."

"Nor yet where we are, here on the rim," Wash remarked; "for I read that the hot water pours over the edge and down the sides of the mound in a perfect cataract."

We waited to hear one more report, which sent another steaming wave of water out to the rim; and then retired to where our tent stood, about a hundred and fifty feet from the foot of the mound. Just as we reached it, a loud explosion made the ground tremble like a drum; and we caught sight of a vast column-head of water over the rim of the basin, and heard it splashing down the farther side of the mound. The steam flew over our heads in great clouds, for the wind was just right to drive it toward our tent; and blew so sharply, that it was all one could do to keep his footing in the teeth of it. We got as much in the lee of the canvas as we could, and stood waiting the display. Nor had we long to wait. At the next explosion, a great shadowy, misty pillar sprang up with a tremendous seething noise,— far up into the sky it seemed. The next instant we were blinded by the sulphurous steam and spray. The strong gale swept the boiling column, and drove all the vapor, and scattered drops down upon us. Halgrim and Weymouth instantly dived back into the tent; but the rest of us, not wishing to lose the sight, ran off to the right along the slope below the mound. But, by so doing, we only got ourselves into greater trouble; for the enormous column, falling back into the basin, overflowed it, and came roaring and dashing down the side of the mound, and was upon us in an instant, foaming like the Oxeara. We had to scamper, and leap from rock to rock; and had our boots well scalded, as well as some uncomfortably hot feet. It ran a perfect deluge all adown the south side of the mound. Beyond the geyser, to the north-east, there is a hollow, or ravine. To that we betook ourselves, and, following it up to the windward of the jet, halted to watch it. The column was no higher than Strokhr, but immensely heavier. The volume of water is far greater. Strokhr consists of scores of rapid *squirts* from the two-foot pipe at the point where the sods choke it, down some thirty feet: Great Geyser, on the contrary, is one massive jet, ten feet in diameter, and rising at a single spring from fifty to seventy feet above the top of the mound. The height of the mound, too (forty feet), adds greatly to the imposing aspect of the column.

"Just observe the form of that jet too!" Raed ex-

7

claimed, still shaking one of his hot feet. "Beautiful, isn't it, as well as grand? Looks like an immense tree, a drooping elm, all white with frost."

"What a mass of water!" said Kit. "That astonishes me most of any thing. It looks just as I have fancied it might otherwise; but I never dreamed there would be so vast a quantity of water ejected at each *puff*. Why, there are more than a hundred hogsheads at every *heave!*"

"It don't stand so steady in the air as Strokhr," Wash observed. "It seems to dance like a candle. The height of the column is increased and decreased each moment with every jet. Strokhr, now, is a bundle of little jets."

"Yes; but there's more water thrown here in one minute than in Strokhr's whole display," Kit remarked.

"What a power there is somewhere at the bottom of that tube," exclaimed Raed, "to hurl up that ten-foot jet with such ease and rapidity!"

"The same power that spouts the lava out of Hecla, Vesuvius, and Ætna," replied Wash; "that drives our locomotives across the continent, and the steamers across the Atlantic, — the expansive power of steam, set in motion by fire-heat; in a word, the power of heat which comes from the pressure of gravitation."

"But this volcanic heat is from the earth's olden fires," Kit remarked: "so, at least, geologists tell us."

"Those geologists who believe the earth was once a blazing star of liquid fire, you mean," corrected Wash. "But they are wrong. The earth was no such a star, according to the best living authorities."

" Among whom, I presume, you mean to include your-self," added Raed maliciously.

" Well," replied the unabashed Wash, "*my theory,* that the earth is the growth of meteors, suggested itself to me with great force and clearness that morning we were on Mount Katahdin. It was based on as careful a computation as *scientists* usually make while figuring on cosmical masses and distances. It came to my mind so forcibly, that I believe it. It seems just as if it must be truth : I *know it must be.* And I don't know why I haven't a right to entertain it, and argue for it fairly. Provided I can sustain it by fair arguments, I don't see why the fact that I am not an old gray-head ought to subject me to ridicule. I don't ask any advantage in argument on account of my youth : therefore I fail to see why I ought to be twitted of being a boy."

This was all said with a candor and self-assertion that would have astonished opposition — had there been any.

Kit exclaimed, " Wash, you are either a great genius, or a very small donkey : I can't just tell which yet. But keep on. Stick to *your theory* till you *prove it:* it will be easier telling then. I don't like to take sides or give an opinion till I hear more about it."

" I don't ask any one to take sides till they are fairly convinced," replied the philosopher. " I haven't had time to work out the theory yet; but I keep it well in mind. Some time, ere many years, I mean to make an effort to let others see this question as I see it."

" In other words, we may look for a treatise on the growth of the earth ? " Raed added in great glee.

"Yes," said Wash, quite regardless of the satire;
"when I have got a few more *respectable* years over my
head. And when you have read it, Mr. Raedway (if you
do me the honor), you will treat the subject with less
levity than you do this morning."

"I have no doubt of it, Wash," concluded Raed.
"I'm not laughing at it now, only to draw you out a
little."

The irruption of the Great Geyser continued about
ten minutes. All the water in a tube seventy-five feet
in depth and ten feet in diameter is thrown out.
Probably, too (as would seem from the extent of the
discharge), the tube connects, by ducts at the bottom,
with other reservoirs of boiling water, which are drawn
upon as the irruption progresses. While the fountain
played, the great basin was white with glittering spray
and foam; and down its sides the hot flood poured in a
vast boiling cataract. The heavy, dull explosions continue during the entire discharge: they are nothing
less than the sudden expansions of steam at the bottom
of the pipe. During the last minute, the jets slacken
and become less frequent, till, with a quick *flip*, all the
water is spouted up, when immediately there is heard
the plunge and gurgle of the returning waters; for, at
the close of an irruption, the basin is left full. The fall
and splashings of the jets are far heavier and more
ponderous than those of Strokhr. As soon as the basin
was sufficiently drained, we entered it, and looked into
the tube. It was already half full again, and the water
rapidly rising.

CHAPTER VI.

WHILE we were looking about the basin, Weymouth came climbing up on the other side.

" What's to be done ? " said he with the air of one who had just witnessed a catastrophe.

" What's to be done with what ? " Kit inquired.

" Why, with the tent and things ! "

We looked over the rim of the basin down to where we had encamped.

" By Jude ! " exclaimed Wash.

" Whew ! " from Raed.

The tent looked as if an elephant had lain down on it, — crushed flat ! Halgrim we could see standing off at some distance, regarding the wreck with rueful looks.

" What did that, Weymouth ? Was it the geyser ? "

" You bet it was ! Hot water came down there like — Old Scratch ! Me and the Icelander was in the tent to keep out of the steam and spray. But, Lord ! pretty quick hot water began to drop by the bucketful.

101

Wind blowed it, ye see. When them big jets flew up,
the gusts would whizzle along, — a perfect sea on't. We
stuck in the tent as long as we dared, and then *dusted*
out and legged it. The Icelander got splashed. Guess
it scalded him some too. He was bare-headed. Had
that big hat of his in his hand when we made our
break. Shouldn't wonder if it killed some of his
lice!"

"Is he lousy, Weymouth?" queried Raed.

"Is he lousy! — why, didn't you know that? Lousy
as a hen! I noticed you didn't give his head quite so
wide a berth as I wanted to when we was all cuddled
down there last night."

"I never thought of his having lice!" Raed exclaimed.

"Why, hain't ye seen him scratch?" demanded the
young tar. "Well, I s'pose ye were eying these 'ere
black rocks too sharp for that. But, bless ye! I knew
he was full of 'em afore we'd got half a mile with him
that fust morning. I never see a fellow *catch* at the
back of his neck, right between his ears, *here* (illustrating it), catch and dig to it, as he does, without he's
chuck-full of 'em."

After this bit of practical information, we turned
our attention to the tent. The strong wind from the
north-east had blown the water over it in such quantities
as to prostrate it completely. The pins on the windward side were pulled out. It lay all in a heap on the
pack-boxes and saddles; but the rubbers had kept the
worst of the deluge off the boxes. The *hay* and great-
coats were well soaked. The milk was washed away;

but the boxes containing the biscuits, sugar, cheese, meat, &c., were comparatively dry — inside. The ground too, all about, was drenched and steaming. Puddles stood in the trenches.

We had plenty of work on hand the rest of the forenoon to move our property. The *hay* had to be shaken up, the boxes and saddles set upon the rocks to dry, and the great-coats, blankets, and tent hung on a line made of the guys, and supported by the poles. Whatever vermin the pedicular Halgrim had introduced among us probably got scalded out in the shower-bath to which the Great Geyser treated our effects.

In the afternoon, before the tent and blankets were dry enough to re-organize, we cut sods again for Strokhr, — about a ton of them. These were piled in a heap around the mouth of the funnel, — piled so near, that a single push would send them toppling down the pipe.

The weather continued cold, cloudy, and windy all day. Against this we fortified strongly with hot soup, coffee, and a fresh supply of milk, cream, and *skyer* from the neighboring *byre*.

By ten o'clock (evening) it came on to rain again heavily. We were still not a little damp *inside* from our drenching of the morning, and, on the whole, passed a rather uncomfortable night. Too much dampness will spoil the hilarity of any party. Wet and dry have a great deal to do with the buoyancy of a fellow's spirits. It is utterly impossible to really relish a joke in a suit of wet clothes, with the rain pouring down.

An accident roused us up toward three in the morning. On a sudden, Strokhr began to growl and rumble.

"Going to heave up, sure as ye live!" said Weymouth, to whom Strokhr's symptoms of nausea had already become familiar.

"Humph!" from Kit in a great state of dissatisfaction: "it will wash away all our sods! No small job to cut a ton of sods!"

We had prepared this dose to give Strokhr at the next irruption of the Great Geyser, to have both perform at once.

Raed got up to look out.

"No wonder he's sick!" he exclaimed. "Why, Kit, half that heap of turf has tumbled into the funnel!"

"Strokhr been taking sods of his own accord!" Wash exclaimed, turning out to reconnoitre.

"That's too confounded bad!" Kit grumbled. "Another hour of good honest digging gone, — lost!"

His sleepy lamentations were cut short by the roar and dashing of the fountain. We went out to see it. Wash improved the occasion to make another sketch from the tent. As Kit had predicted, all the remaining sods were swept in with the water. The first "heave-up" was immediately followed by a second, and this by a third, and so on during the greater part of the forenoon. We took occasion to measure the height of the column, as has been previously mentioned. This irruption also settled a question which had been somewhat warmly discussed by Raed and Kit. Raed did not believe that the quantity of sods thrown into the funnel had much to do with the violence of the irruption: he contended that a hundred-weight of sods would induce as high a jet as a ton. Kit argued that the force of the jets would be somewhat in proportion to the weight of sods. The discharge, on this occasion, was seemingly far less violent than that produced by the two tons of sods on the evening of our arrival.

The rain continued all day. We began to wish that we had been content with G. G.'s first irruption, and gone back to Reykjavik. The cold and dampness were extremely uncomfortable. We had all taken "colds," and had "sneezing-matches;" object being to see which could draw the loudest echo from the side of the Langarfjal.

The next day (June 15) was fair.

More sods were cut for Strokhr, — another ton, — and heaped about the orifice.

G. G. evinced no symptoms of another irruption. Kit made a sketch of the peaks of Hecla, towering grandly to the southward at a distance of twelve or thirteen miles.

The night was chilling, for a June night; though the sun was scarcely out of sight ten minutes behind the mountains. More "sneezing-matches" next morning.

A little after ten, G. G. fired his first audible gun, seemingly ten miles down in his abdomen. Wash first noticed it. For the next hour, they drew gradually *nearer* and *heavier*. By eleven, they were let off with startling loudness. Then came a lull. For over an hour there wasn't a sound, save from an occasional discharge of bubbles up the funnel. "False alarm," Wash had pronounced; but at twenty minutes past

twelve the guns began again, and by one the explosions
began to *bulge up* the water into the basin.

"Time to give Strokhr his dose!" cried Raed.

He and Kit ran down, and with the spades thrust
over the heaps of clods. We then all clambered up the
side of the bluff above the *hvers* to get a good view of
the expected display. Strokhr began to growl and
rumble; but, before he had got sick enough to *heave,*
the Great Geyser went up in all its majesty. Oh, it
was grand! We had eyes for nothing else. That
booming, plunging, roaring pillar of steam and glittering spray — 'twas worth waiting for! I never expect
to see a more magnificent spectacle. Strokhr began
when Great Geyser had been playing about five
minutes. The poor little *churn* cuts but a sorry figure
beside its mighty relative. The superior height of the
Great-Geyser site and its lofty mound, altogether, make
it vastly overtop Strokhr. The far greater volume of
its waters, too, renders the latter still more insignificant.
Nevertheless, the black, inky jets of Strokhr are in
startling contrast with the *white* and sparkling geyser-
pillar. They need to be seen together; and we would
respectfully advise tourists to secure a joint performance, if they can in any way manage it.

As many of our readers may, like ourselves, feel in
terested in the *whys* and *wherefores* of these tremen
dous irruptive fountains, we subjoin Prof. Bünsen's
"Theory of the Geysers" as set forth by Prof. John
Tyndall, believing it to be the latest and best authority
on a subject which has heretofore perplexed scientific
men not a little to explain.

A Theory of the Geysers.

In his third lecture on heat, Prof. Tyndall says, " I must now direct your attention to a natural steam-engine, which long held a place among the wonders of the world. I allude to the Great Geyser of Iceland. The surface of Iceland gradually slopes from the coast towards the centre, where the general level is about two thousand feet above the sea. On this, as a pedestal, are planted the *jökull*, or icy mountains, which extend both ways in a north-easterly direction. Along this chain occur the active volcanoes of the island; and the thermal springs follow the same general direction. From the ridges and chasms which diverge from the mountains enormous masses of steam issue at intervals, hissing and roaring; and, when the escape occurs at the mouth of a cavern, the resonance of the cave often raises the sound to the loudness of thunder. Lower down, in the more porous strata, we have smoking mud-pools, where a repulsive blue-black aluminous paste is boiled, rising at times in huge bubbles, which, on bursting, scatter their slimy spray to a height of fifteen or twenty feet. From the bases of the hills, upwards, extend the glaciers; and above these are the snow-fields, which crown the summits. From the arches and fissures of the glaciers vast masses of water issue, falling at times in cascades over walls of ice, and spreading for miles over the country before they find definite outlet. Extensive morasses are thus formed, which add their comfortless monotony to the dismal scene already before the traveller's eye.

Intercepted by the cracks and fissures of the land, a
portion of this water finds its way to the heated rock
underneath; and here, meeting with the volcanic gases
which traverse these underground regions, both travel
together, to issue, at the first convenient opportunity,
ither as an eruption of steam, or a boiling spring.

"The most famous of these springs is the Great Geyser.
It consists of a tube seventy-four feet deep, and ten feet
in diameter. The tube is surmounted by a basin, which
measures from north to south fifty-two feet across, and
from east to west sixty feet. The interior of the tube
and basin is coated with a beautiful, smooth, siliceous
plaster, so hard as to resist the blows of a hammer; and
the first question is, How was this wonderful tube con-
structed? how was this perfect plaster laid on? Chemical analysis shows that the water holds silica in solution;
and the conjecture might therefore arise, that the water
had deposited the silica against the sides of the tube and
basin. But this is not the case. The water deposits no
sediments : no matter how long it may be kept, no solid
substance is separated from it. It may be bottled up
and preserved for years as clear as crystal, without showing the slightest tendency to form a precipitate. To
answer the question in this way would, moreover, assume
that the shaft was formed by some foreign agency, and
that the water merely lined it. The geyser-basin, however, rests upon the summit of a mound about forty feet
high; and it is evident, from mere inspection, that the
mound has been deposited by the geyser. But, in building up this mound, the spring must have formed the
tube which perforates the mound; and hence the con-

clusion that the geyser is the architect of its own tube.

"If we place a quantity of the geyser-water in an evaporating-basin, the following takes place: In the centre of the basin the liquid deposits nothing; but at the sides, where it is drawn up by capillary attraction, and thu subjected to speedy evaporation, we find silica deposited. Round the edge, a ring of silica is laid on; and not until the evaporation has continued a considerable time do we find the slightest turbidity in the middle of the water. This experiment is the microscopic representant of what occurs in Iceland. Imagine the case of a simple thermal, siliceous spring, whose waters trickle down a gentle incline: the water thus exposed evaporates speedily, and silica is deposited. This deposit gradually elevates the side over which the water passes, until finally the latter has to take another course. The same takes place here: the ground is elevated as before, and the spring has to move forward. Thus it is compelled to travel round and round, discharging its silica, and deepening the shaft in which it dwells, until finally, in the course of ages, the simple spring has produced that wonderful apparatus which has so long puzzled and astonished both the traveller and the philosopher.

"Previous to an eruption, both the tube and basin are filled with hot water: detonations which shake the ground are heard at intervals, and each is succeeded by a violent agitation of the water in the basin. The water in the pipe is lifted up so as to form an eminence in the middle of the basin, and an overflow is the consequence. These detonations are evidently due to the production

of steam in the ducts which feed the geyser-tube; which
steam, escaping into the cooler water of the tube, is there
suddenly condensed, and produces the explosions. Prof.
Bünsen succeeded in determining the temperature of
the geyser-tube, from top to bottom, a few minutes before
a great eruption; and these observations revealed the
extraordinary fact, that at no part of the tube did the
water reach its boiling-point. In the annexed sketch
I have given on one side the temperatures actually
observed, and on the other side the temperatures
at which water would boil, taking into account both the
pressure of the atmosphere and the pressure of the super-
incumbent column of water. The nearest approach to
the boiling-point is at A, a height of thirty feet from the
bottom; but even here the water is two degrees Centigrade, or more than three and a half degrees Fahrenheit,
below the temperature at which it could boil. How,
then, is it possible that an eruption could occur under
such circumstances?

"Fix your attention upon the water at the point A,
where the temperature is within two degrees Centigrade
of the boiling-point. Call to mind the lifting of the column when the detonations are heard. Let us suppose,
that, by the entrance of steam from the ducts near the
bottom of the tube, the geyser-column is elevated six feet,
— a height quite within the limits of actual observation:
the water at A is thereby transferred to B. Its boiling-
point at A is 123.8°, and its actual temperature 121.8°;
but at B its boiling-point is only 120.8. Hence, when
transferred from A to B, the heat which it possesses is
in excess of that necessary to make it boil. This excess

of heat is instantly applied to the generation of steam:
the column is thus lifted higher, and the water below is
further relieved.

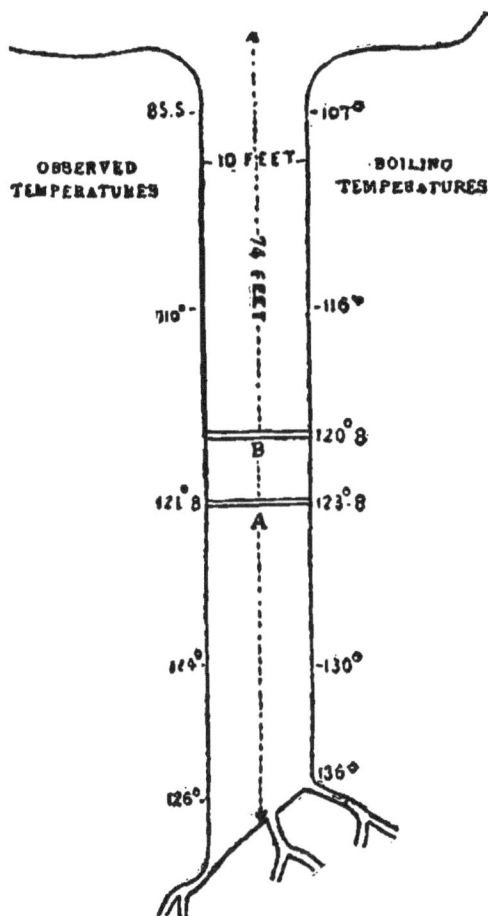

OBSERVED
TEMPERATURES

BOILING
TEMPERATURES

85.5-

-107°

-10 FEET-

-74 FEET-

710°-

-116°

120°8
B

121°8

123°8
A

114°

-130°

126°.

136°

"More steam is generated; from the middle, down-
wards, the mass suddenly bursts into ebullition; the

water above, mixed with steam-clouds, is projected into the atmosphere; and we have the geyser-eruption in all its grandeur.

"By its contact with the air the water is cooled, falls back into the basin, partially refills the tube, in which it gradually rises, and finally fills the basin as before. Detonations are heard at intervals, and risings of the water in the basin. These are so many futile attempts at an eruption; for not until the water in the tube comes sufficiently near its boiling temperature to make the lifting of the column effective can we have a true eruption.

"To Bünsen we owe this beautiful theory. And now let us try to justify it by experiment. Here is a tube of galvanized iron, six feet long, AB, and surmounted by this basin, CD.

"It is heated by a fire underneath; and, to imitate as far as possible the condition of the geyser, I have encircled the tube by a second fire, F, at a height of two feet from the bottom. Doubtless the high temperature of the water at the corresponding part of the geyser-tube is due to a local action of the heated rocks. I fill the tube with water, which gradually becomes heated; and regularly, every five minutes, the water is ejected from the tube into the atmosphere.

"But there is another famous spring in Iceland, called the Strokhr, which is usually forced to explode by stopping its mouth with clods. We can imitate the action of this spring by stopping the mouth of our tube, AB, with a cork. The steam below will finally attain sufficient tension to eject the cork; and the water, suddenly relieved from its pressure, will burst forth in the

atmosphere. In the following figure I have given a
section of the Strokhr: —

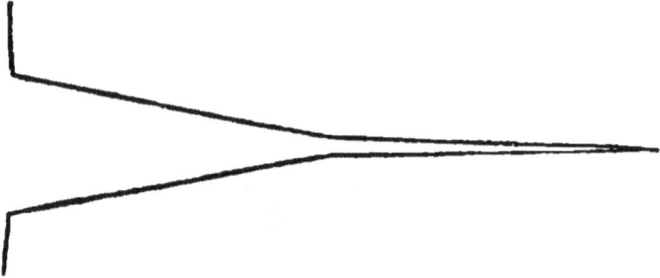

" By stopping the tube with corks through which tubes
of various lengths and widths pass, the action of many
of the other eruptive springs may be accurately imitated.
Here, for example, I have an intermittent action: dis-
charges of water and impetuous steam-gushes follow
each other in quick succession, the water being squirted
in jets fifteen or twenty feet high. Thus it is proved,
experimentally, that the geyser-tube itself is the sufficient cause of the eruptions; and we are relieved from
the necessity of imagining underground caverns filled
with water and steam, which were formerly regarded as
necessary to the production of these wonderful phe-
nomena.

" A moment's reflection will suggest to us that there
must be a limit to the operations of the geyser. When the

tube has reached such an altitude that the water in the depths below, owing to the increased pressure, cannot attain its boiling-point, the eruptions, of necessity, cease. The spring, however, continues to deposit its silica, and often forms a *laug*, or cistern.   Some of those in Iceland are forty feet deep.   Their beauty, according to Bünsen, is indescribable.   Over the surface curls a light vapor: the water is of the purest azure, and tints with its lovely hue the fantastic incrustations on the cistern walls; while at the bottom is often seen the mouth of the once mighty geyser.   There are in Iceland vast, but now extinct, geyser-operations.   Mounds are observed whose shafts are filled with rubbish, the water having forced a passage underneath, and retired to other scenes of action. We have, in fact, the geyser in its youth, manhood, old age, and death, here presented to us, — in its youth as a simple thermal spring, in its manhood as the eruptive column, in its old age as the tranquil *laug*, while its death is recorded by the ruined shaft and mound which testify the fact of its once active existence."

CHAPTER VII.

ON the morning of the 17th of June we bade farewell to the geysers, and, with our long train of
ponies, went galloping back toward Reykjavik. Kit had
thrown a few parting sods into Strokhr's pipe, for which
that irascible old gentleman growled and groaned us out
of hearing; and just as we were turning the angle of the
ridge, before passing the shepherd's *byre*, we saw him
spouting up the black contents of his ever-bilious stomach high over the intervening knolls.

Our journey back to Reykjavik was over the same
trail by which we came out. I shall not venture a
second account of its scenery. We arrived in town at
a few minutes after one on the afternoon of the 18th.
What to do with our horses was a somewhat puzzling
question; for, as will probably be remembered, we had
been obliged to purchase eleven outright. After some
discussion, it was decided to have an auction-sale, with
Halgrim for auctioneer. Accordingly, after a hearty

116

dinner at the hotel, we hired three "loafers" to go out and notify the public generally of the proposed "vendue." In the course of an hour, there was a crowd of from seventy-five to a hundred; and we directed Halgrim to proceed without further delay. Auction-sales, it appeared, were not unknown in Reykjavik. Halgrim perfectly comprehended the process, and went into it with a will. Nevertheless, the Reykjavik public fought very shy of us. I suspect that disastrous conspiracies were entered into among the jockeys to keep down the price. Nags for which we had paid from forty to fifty rix-dollars, we could now only get nineteen offered for. They looked as well, too, as when we started off, for aught we could see. I shall always think the Reykjavikers took a mean advantage of our necessity on that horse-auction occasion. The auctioneer, apparently, did his best; but, for all that, those eleven horses only brought two hundred and twenty-three rix-dollars, — about one hundred and thirteen American money.

Halgrim was then paid, with a handsome little gratuity. We shook hands and bade him good-by (we didn't kiss him), and then went down to the jetty. There lay "The Curlew," looking, for all the world, like some dear old friend. We raised a cheer, which soon brought off the boat with Hobbs and Bonney. Capt. Mazard was awaiting us with a hearty hand-grasp over the rail to help us aboard.

" How are you, boys ? "

" Cap., how do you do? " and Kit sings out, " Captain, how about the *cntman's* daughter, the *Skén Jomf-u?* What success at the stone house ? "

The captain shakes his head very sagely and gravely.

"Raed, my boy," beginning with great solemnity, "I hope you will try to bear it calmly."

"I shall try to," replies Raed, showing symptoms of internal agitation. "I hope you will try to do the same on your own part."

"'There is serious reason to apprehend," resumed the captain with a melancholy bow, "that the affections of the *Skén Jomfru* are — can you bear it, Raed? — are already engaged."

Raed staggers back with clinched fists. The captain gazes gloomily and suicidally over into the water. Thus ends the tableau; and thus terminates all possibility of a romance at Reykjavik.

Palmleaf's cookery had a decidedly good taste after eight days of dry biscuit and soups. We had a "grand feed" that night. The darky was fain to exclaim that he "neber seed sich appletites as dat gazer-water did make."

According to our pre-arranged plan, we sailed the next morning for Akureyri on the Eyja Fiord, the second largest town of Iceland. Reykjavik is situated, as will be seen from the map, in nearly the south-west corner of the island. Akureyri, on the contrary, is in almost the north corner. We desired to visit the region of the more recent volcanic irruptions to the north-east of the Eyja Fiord, as also the Godafoss and Dettifoss cataracts.

To reach Akureyri from Reykjavik, we sailed around the north-west coast of the island, and, proceeding along the northern coast, descended the Eyja Fiord.

This route is practicable only during those seasons

when the passage between Iceland and Greenland is comparatively free from those vast ice-fields which periodically obstruct it, causing what are known as the "cold summers" in Iceland. This same cause also affects the summers as far south as England and Ireland.

Sailing out of the port of Reykjavik, we doubled the Snæfel promontory on the upper side of the Faxa Fiord by ten o'clock in the afternoon; and, keeping well out to sea, passed the mountainous northern capes of the island on the morning of the 20th. The sun did not set at all that night. At midnight it stood a fair handbreadth above the sea. A strange place this, — the sun shining in the north at midnight! The ocean was calm: the schooner made but little headway. Great icebergs drove slowly southward, white as marble, and often shining like silver in the waning sun.

Thence our course was almost due east.

On the 21st we passed the entrance to the *hunaflöi* (bay). During the night, if thus it can be termed, of the 22d, we sailed athwart the embouchure of the Skaja Fiord; and, toward noon of the 23d, entered the Eyja Fiord, — a long, narrow gulf, walled on both coasts by high, bare mountains capped with snow. The fiord itself seems to occupy a mighty rent in the island, extending far down toward its centre. At the foot of this gulf is Akureyri. The wind was fitful and light; and it was not till near midnight that we reached the quaint little northern hamlet. From the mouth of the fiord to Akureyri is a distance of about forty miles, so deeply does this narrow bay penetrate the island.

Of the scenery I gained very little idea at the time.

A bad cold, taken from the damp clothes at the geysers, had at length thrown me into a sort of fever, which, during the last two days of the voyage, had prostrated me completely.   But for the energetic sweats and fever-teas with which my comrades tortured me, I should have died, I make no doubt.   It was five days after reaching Akureyri before I went on deck.   Of my own chagrin at thus being the cause of so much delay I have no need to speak.   During this time, however, the other boys had made the acquaintance of nearly every-body in town, and, among the rest, of an intelligent young Dane, named Havisteen, of about our own age He was either the son or nephew of a merchant whc makes this place his temporary residence.   His home was, as I understood, at Copenhagen.   Young Havisteen could speak some English, and understood Icelandic enough to converse to a considerable extent with the people who trade at the little seaport.   He was also an amateur geologist in his way, and had been as far as Reykjalith on an expedition of his own.   The boys found him a very congenial spirit.   He was knocking about without much to do, and jumped at the chance of joining our party on any sort of a trip.   Falling in with him was a streak of good luck, — so Kit thought. "He was such a jolly good fellow, he would do just as well as any guide we could get; and, better still, he had two horses of his own, and could hire half a dozen more of town-people he knew."

All this, and plenty more, I heard from the boys, who were going back and forth from the town to the schooner all the time.   They kept ringing it in my ears as I lay

AKUREYRI.

there feebly on my back. How it does tire an invalid to hear and see a hale, well person talk and tear round! and, *vice versa,* I suspect that the invalid is about the biggest sort of a *bore* to the well one. Long before I had got back *vim* enough to stumble up the companion-way, they had every thing arranged, and were only wait-ing for my *restoration.*

My first intelligible view of Akureyri was on the morning of the 29th of June. If I had been attended by a physician, I dare say he would have positively for-bidden me to leave the yacht. Capt. Mazard said all he could to have me remain with him; but Kit was urgent for me not to lose the "fire region" about the Namarfjal. I didn't like much to stay behind; and, knowing how impatient the boys were getting, I mustered my strength, and started out, notwithstanding the captain's ominous prediction that I "was a dead man" to do it.

Akureyri is a row of tarred wooden shanties, with two or three stone houses, extending along the beach at the foot of a steep ridge, or bluff, which rises abruptly in their rear to the height of three hundred feet, and thence slopes off to the summit of the Wind Jokuls three thousand feet in height. The fiord is here about a mile in width, and extends for some little distance farther up the valley, which makes back into the island, between two high mountain-ranges. The hamlet is said to contain a population of a thousand inhabitants; but, from its size, this seemed to me clearly impossible. The space between the hill and the bay is so narrow, that the doorsteps of the houses are actually on the very beach. One can almost step from a boat into the houses.

It is at this port that all the trade of Northern Iceland goes on ; and, if what we saw was a fair specimen
of its usual business, I should account it a very small
trade indeed.

Young Havisteen, with the horses and baggage, was
awaiting us as we rowed ashore from the schooner ; and
I at once had the honor of an introduction. He is a
good-natured, and, withal, a very good-looking young
fellow ; lively, smart (for a Dane), and in possession of
a good deal of practical information on this, that, and
the other subject. His tastes, likes, and dislikes — so far
as we could discover through the somewhat clumsy
splicings of the two languages which we made — were
much the same as our own ; and, as a natural consequence, we *cottoned* to him amazingly. It seemed *real
good*, as well as novel, to find a chap who thought as we
did, away up here in this out-the-way corner of the
world. The only obstacle to a perfect understanding
was the trouble of making out what he said. Oh, what
a glorious thing it will be when all nations speak one
language, — one universal language which everybody
can understand ! God speed that time, I say ! What a
bother of lexicons and grammars, and " new methods,"
and teachers, and years of study, it will do away with !

We had five horses to ride, and two to carry our baggage.

Young Havisteen (his Christian name was Jan, Danish
for John, which first Kit, and then all the rest of us, had
come to call him before night) didn't believe that extra
saddle-horses were necessary ; therein differing from
the nefarious Reykjavikers. He declared, moreover. that

two horses could just as well carry our luggage as five. It should be explained, however, that the north-country horses are rather larger than those we bought for our geyser expedition. The northern portions of the island far excel the tract about Reykjavik for agricultural purposes. The average temperature at Akureyri, through the year, is 32° Fahrenheit, so young Havisteen told us; but during the winter season it is sometimes as low as — 32°.

Riding out of the town, we continued down the beach on the west side of the fiord to the head of it, where we forded the Eyja River, which there makes in from the valley. It has seven mouths and a very respectable delta. When the tide is in, it is half a mile in width. This statement has a very large sound, which I may qualify by adding that the stream itself is not larger than the Charles River at Boston, or the Ashley at Charleston. Its resemblance to the ancient Nile, as young Havisteen very naively remarked, ends at a very short distance from the sea.

Passing the Eyja, we turned to the north-east, and entering a *skarth*, or mountain hollow, ascended by it toward the summit of the jokul ridge on the east side of the fiord. We were two or three hours getting over this. Near the top we encountered a smart snow-squall, which rapidly whitened the dark ledges. Ice lay in all the little hollows. It was a bleak place. We gladly descended into a vale beyond, at the bottom of which whirls a deep, rapid stream. Its name, the young Dane informed us, was spelled F-n-j-o-s-k-a. We all gave *in* on its pronunciation. Kit suggested that the first syllable might be approximated by blowing one's nose sharply. Raed thought that an ordinary hiccough might well stand for the last two. The trail led directly into it; but it was not without some misgivings that we undertook the ford, — a perfect rapid, foaming among bowlders over a rough bottom strewn with huge, round, slippery pebbles, among which our horses crippled and sprawled. Despite our attempts at drawing up our feet, we all got our boots full of water; and Wash's pony, getting out of his depth, actually swam under him, and only came to the bank some fifty yards below.

Oh! Iceland never can advance in civilization, nor any thing else, till they have these terrific torrents bridged. That was very evident to every one of us on getting out of the Fnjoska.

Our route next led through what came nearest to a forest of any thing we saw in the country. For a long way, birch-shrubs, some of them ten and fifteen feet in height, stood thickly on both sides. The leaves were now just unfolding. It seemed decidedly home-like; and, in consideration of my weakened frame, the boys kindly consented to encamp here for the remainder of the afternoon and night. The tent was pitched, and the blankets spread on a *springy* heap of birch-twigs. Once out of the saddle, I *turned in* on them, and, swaddled up in the great-coats, went to sleep like a year-old baby.

Meanwhile, as was afterwards told me, the boys, in collecting fuel for a fire, started a "blue fox," and had a prodigious race after him, — one of the "best things," Kit said, he ever saw. After a very exciting hunt, the game was *holed* about a mile away; and, by tearing off the

AN ICELAND FOX.

rubbish, they contrived to shoot him: all of which I remained in blissful ignorance of till they showed me the skin some hours after. Those who have observed the peculiar hue of a genuine Maltese cat will have a very good idea of the color of an Iceland fox, which is a species quite unlike the gray and red Reynards of any other country. Its claws are nearly as sharp and retractile as those of a cat. Raed told me that this one was not more than two-thirds as large as the red fox of New England.

Our provisions were much the same as on our geyser trip. Supper was prepared at eight o'clock. I did very little but sleep till late the next morning; when, after a breakfast of thick soup, we again set out. A gallop of an hour took us past the church and station of Hals; and, entering the Lijosavatn Skarth (Light-water Lake Ravine), we soon came out upon the Lijosavatn, — a pale sheet of water, in which patches of ice were still floating. Two loons were sailing about. Their hollow, quavering cry strangely reminded us of the Maine lakes, where we had so often heard them while "prospecting" on the Katahdin ridge. Winding over the mountains beyond the lake, we descended into the valley of the Great *Skjalfandi-fljot.* Sneeze, reader, and you will have the first word: the second is simply *flot:* all together, the phrase signifies "shimmering flood." The river is thus named from the quivering motion of its waves. Wallowing through a fen, we approached the bank, and hailed the ferryman, who presently took us across. The *Skjalfandi* is the second river of Iceland, and far too deep and broad to be forded. Wash compared it to the Merrimack in size. I am uncertain as to the correctness of this comparison. It seemed as large, though it resembles that industrious river in no other particular. The ferryman's name, on this occasion, was Páll. Young Havisteen wondered whether he was aware of the associations it would call up in certain purlieus of London or New York.

The Godafoss cataract is about a mile above the ferry. I did not feel able to accompany the other boys up to it. While they were gone, I took a nap in the old ferry-boat. Raed tells me that it is an almost perfect Niagara in miniature; resembling our world-wide wonder in its rapids, its Goat Island, its Horse-shoe Island, and many other particulars. The lava-cliffs which enclose the river for some distance below the falls are, he says, of wonderful height, and almost sooty blackness.

Climbing up from the river, we rode on for ten or a dozen miles over a rough, bare country, with distant pale-blue jokuls set on the horizon.

At about four o'clock (afternoon) we passed a small church and farm, which our Danish friend called *Thverá;* and crossed a small stream known as the *Laxa,* or Salmon River.

An hour later a fine broad expanse opened to view from a hill-top.

"What *vatn* is that?" Kit asked.

"*Myvatn,*" replied Jan. "*My* means *midge* in your language: so there you are, Midge Lake. And there is the farm and church of Reykjalith out past the head of the *vatn,* — three English miles, just about."

"Had we best go to the farm, Jan?" said Raed doubtfully.

"*Ja.* Good old chap, the farmer. Always takes in tourists. His wife is a pleasant woman. She makes good *skyer* and nice cakes. We shall have the *badstöve* all to ourselves."

" The bad what ? " exclaimed Kit.

" The *badstöve :* that's the guest-room, or rather bed-room," Jan explained.

A long way below the *byre* we were met by a shaggy house-dog, — a terribly-ferocious little chap, all *yap* and bristle. Had his size been in any proportion to his ferocity, we should have been torn piecemeal on the spot. Three ravens too, which had been perched leisurely on the gables of the hovels, came flapping out, and, wheeling tamely about our heads, kept settling down in the path, croaking dismally when the approach of the ponies made it necessary for them to hop aside. They strongly resemble our American crows in size, and in the shape of their large, knowing heads. Wash pronounced them rather larger. They plainly belong to the same order of birds. At sight of them, young Havisteen repeated a verse of an Icelandic poem entitled "Hrafna" ("The Raven"); and afterwards gave us its English translation, as follows : —

> " Raven sits on gable-tree ;
>   Watch! Death is onward creeping :
> Short the life of him will be
>   Who 'neath that roof lies sleeping."

From which it would appear, that in Iceland, as in other lands, the raven has been regarded as a bird of ill omen.

The terrific barks of the dog gave timely notice of our coming in-doors. As we rode up, both the old farmer in his skull-cap, and his wife in a tall white turban, stood at the door with kindly faces. Behind them smiled a cherry-cheeked lass in a tight black jacket (?); while at another doorway farther along the establishment stood a tall, gaunt youth of twenty, or thereabouts, in wonderfully-tight pants and a leathern frock. Another dog — that had probably up to this moment been slothfully asleep — rushed out; and together they nearly deafened us, rendering all attempts at vocal salutation on our part an utter failure. But it was a pleasure for us to observe that the family had all recognized Havisteen. We could *see* their mouths going, and catch now and then a sound over the canine uproar. All three seemed to be bidding him and us — as we judged from kind looks in our direction — a hearty welcome. Just at this juncture, another dog — a puppy — bobbed out past the old lady's skirt, and began on a very sharp key. Havisteen had now fairly to shout; so did the old lady; so did the old farmer. Yet it seemed not to occur to either of them that any thing was the matter. Finally, the young fellow, as if struck by the idea that so much *doggerel* was a sort of old-fashioned nuisance, which he, as the representative of the rising generation, ought to discountenance, stole quietly out, and, with a couple of sly cuts from a whip he held, changed the boisterous barking into a chorus of surprised yelps. The old lady stopped short, and stared at him in great astonishment and disapproval. The youngster slunk back toward his doorway. He didn't venture to cut at the pup.

AN ICELAND BYRE.

Jan now dismounted, and bade us do the same. The farmer and his son took off the saddles, and led the ponies away to their pasture. The old lady conducted us *under* a low doorway. The cherry-cheeked lass disappeared in the darkness of a long passage, which we rendered still darker by entering; the door behind us being its only window. This was the main hall, off which blanketed doorways branch into the various apartments, or rather houses; for each apartment is one entire house. Putting up our hands, we groped on after the good-wife, occasionally grazing against the lava-walls of the passage. After going, as I judge, from thirty to forty feet, and having made several *turns* and *bends*, a blanket was pulled aside, and we were ushered into the *badstöve.* It had a window of glass, — two windows, in fact; though the second was a very diminutive one. Whether the floor was of clay, or a kind of "Nicholson pavement," I never could fully decide. There were several wooden chairs and two beds. The beds would, I fancy, have astonished a New-England matron. The bedsteads were simply huge boxes' on legs, into which two exceedingly corpulent beds of eider-down are tumbled. On top of these feathery mountains are piled coverlets, puffs, and spreads *ad infinitum.*

"By Jude! here's a soft thing," was Wash's exclamation on giving one of them an experimental poke.

His next query was, "Wonder if there are any bugs in 'em."

To the credit of the housewife, I should here affirm:

that we found but *two bugs;* and they were seemingly *seed ones* that had long since retired from active
life.   The boxes containing our biscuits, coffee, cheese,
meat, &c., were brought in, and given in charge of the
women, with the request to draw on them for soups,
and add whatever they chose from their own supplies at
our expense.   The result was a steaming supper of excellent coffee, soup, skyer, biscuit and butter, cheese,
and some delicious little cakes made of native wild
corn and cinnamon.   Seated around a table in the
badstöve, with the luxury of a white cotton tablecloth,
we enjoyed it amazingly, — all the more that it had
come on to rain, pattering softly on the turf-covered
roof over our heads.   Young Havisteen, at our request,
gave the farmer and his family an invitation to take their
tea with us; but, for some reason, they declined to do
so.   I had become too much exhausted to long resist
the charms of those downy beds; though they were so
soft as to actually frighten me on first getting in.   So
violent was the contrast between these feathery depths
and the birchen "shake-down" of the preceding evening, that it was really some little time before I could
reconcile myself to such extravagant softness ;   and,
after going to sleep, I kept jumping up from dreams of
sinking in bottomless *laugs* for more than an hour.

   The rest of the party didn't come to bed till a late
hour; and I learned next morning that they had been
out into the kitchen, ostensibly after a dram of milk,
but really to get another glimpse of the cherry-cheeked
lass in the black jacket.   Kit also informed me that

Wash (he seemed greatly scandalized and disgusted with him) was carrying on a most ridiculous and altogether senseless flirtation with her: he felt sure he would disgrace us all yet with his ineradicable *bent* toward nonsense.

CHAPTER VIII.

FOR breakfast we had boiled mutton in the place of soup; otherwise the programme was unvaried The rain still continued. It was a dark, drizzling morning. We were glad to be in so comfortable quarters, and decided to let the *Namarfjal* severely alone till the weather faired. During the forenoon we visited the church, which is near the farm. It is quite an interesting edifice — for the place. It contains a font-basin for baptisms; and in the chancel there is a curiously-colored lantern. There were three chasubles, or robes for the priest, — one of white silk, and very old; one of red damask; and the third of crimson velvet. We saw also a curious wrought-iron candle-rack; but a far more interesting object to us was a bronze ring at the door, said to have been taken from the ancient heathen temple which stood here in the days when the Icelanders worshipped Thor and Odin. This was the ring which they used to dip in the blood of the victims, and make their vows on. We could but wonder at so san-

132

guinary a memento ever getting a place, even at the door of a Catholic church;* for, in consideration of the purpose it had served, it could but be regarded as "a pretty hot old relic" (I quote from the words of the scandalous Wash).

There being nobody about who would be likely to be shocked, Kit put on the red-velvet chasuble, and read a long Latin service, prayer, or oration, — none of us knew exactly which, — from a very dilapidated volume, which lay in close company with the pulpit Scriptures. This was merely to try the effect of the red velvet and the ritualism; not out of any rowdyish contempt for the church or its belongings.

After much urging, we persuaded the old farmer to take dinner at our table. When we were well seated, and comfortably disposing of our soup, Raed said, "Ask him, Jan, whether he has got any *sagas*."

Jan did so in Icelandic.

"*Ja*," he had; and a light broke over his broad honest face, and shone in his gray-blue eyes, that we had not before seen.

"What *sagas* has he got?"

"He says," replied Jan, "that he has the *Saga af Asmundr*, *Viking*, and the *Gretla Saga*, and the *Saga af Ambrosia ok Rosamanda*."

"Ask him, Jan," said I, "if he will read some from the *sagas* to us."

"But you couldn't understand a word of it," Jan objected.

"Couldn't you tell us what it was about, pretty

* Now Lutheran.

nearly, after he had read a paragraph?" queried
Raed.

Jan reflected a moment.

"*Ja*, I think I could, — the most of it."

"Will you do it?" Kit asked.

"Oh, *ja!*" if we said so. And he immediately asked
the old Icelander if he would read from the *sagas* to
the young Americans. At first, the man looked rather
wonderingly at us; but, on the young Dane explaining
to him how we had planned to understand it, he seemed
pleased, and at once assented.

But which *saga* should he read to us, so Jan inter-
preted for him.

Of course we had very little choice, for a very good
reason.

"Tell him to read us the one he likes best," said
Raed.

" He says," Jan informed us, " that he likes the *Gretla* best; and he thinks we shall like it best too; for
it is about as brave a lad as ever left his native
shores."

We bowed our acknowledgments to the old man for
the implied compliment thus happily expressed, and
bade Jan tell him we *knew* we should like the *Gretla
Saga*.

As soon as we had finished dinner, the old farmer
went out after his beloved *Gretla*. Young Havisteen followed him, to ask his wife and the young folks to come in ;
and, when they found it was a *saga*-reading, they smil-
ingly accepted our invitation, and edged pleasantly in
behind the venerable reader. It was truly a "quaint

and curious volume of forgotten lore" which the old Icelander brought forward. The leather cover was black with age; not a word in it printed; all written with a pen on coarse yellowed paper. As fast as a *saga* wears out, the good people recopy it with great care and patience.

The farmer seated himself at the table, which had, meanwhile, been cleared of its platters and plates; and carefully adjusting a pair of spectacles with round silver bows, very heavy and quaint, he opened the time-worn book, remarking to Jan, which Jan remarked to us, that, as we might not perhaps be able to hear it all if he were to read in course, he would read parts such as he deemed most interesting. To this we of course assented with respectful bows. Never shall I forget the kind of affectionate reverence with which the old man began to read. Insensibly we all felt a sort of sabbath sensation stealing over us, Wash only evincing a sacrilegious propensity to hitch his chair along a few inches farther toward the cherry-cheeked lass of the black bodice. We all frowned at him industriously, particularly Kit. The good-wife had brought her knitting; the shaggy house-dog walked in, and sat down demurely; the pup followed shortly after, and didn't even so much as chase his tail round once, but, with a glance at the old dog, sat down likewise. After a number of *ahems*, each of which I kept confidently presuming to be the final one, the Icelander began to read. His tone was of that *worshipful* modulation which we so often hear very pious persons make use of when reading the Scriptures. He read for three or four min-

utes, then paused for young Havisteen to translate to us.
From the first, it had occurred to me to secure the
translation, if possible, to go with my narrative of our
Iceland tour. Accordingly, after Jan had repeated in
substance what the farmer had read, and while he was
reading a second paragraph, I caught down in my
diary (writing recklessly over dates and days) the
main part of the story as Jan had translated it. Afterwards, during our homeward voyage, I wrote it out more
in full, and now submit it to the reader — not as a lite-
ral translation, but as a liberal paraphrase — in parts
as the old man read them to us.

[As young Additon's translation was, from the nature
of the case, but a "liberal paraphrase," we have deemed
it better, on the whole, to substitute for it a *more careful
rendering of this beautiful story*, which we take from a
well-known English translation of the Gretla. — Ed.]

GRETTIR THE STRONG.

One morning, after a night of storm on the coast of
Norway, the servants ran into the hall of a wealthy
bonder named Thorfin, to inform him that during the
night a ship had been wrecked off the coast, and that
the crew and passengers were congregated on a neighboring sandy holm, signalling for help. Up started the
bonder, and hastened to the strand: he ran out a large
punt from his boat-house, and, jumping in with hi
thralls, rowed lustily to the rescue. The shipwrecked
people belonged to a merchant-vessel from Iceland, which
had been driven among the breakers during the darkness,

and had g..ne to pieces; yet not before a portion of the lading had been brought ashore.

Among the shivering beings gathered on the sand-strip was Grettir, the son of an Icelandic chief who lived at Bjarg, in the middle frith. He was then a young man, tall and muscular, with large blue eyes, bushy hair, and a freckled face.

Thorfin received the half-frozen wretches on board his boat, and rowed them to the mainland; after which he returned to the holm, and brought off the wares. In the mean time, the good housewife had been lighting fires, preparing beds, routing out dry suits, and making hot ale ready for the sufferers; and right gladly they were treated, you may be sure.

Well, the chapmen staid a week at the farm whilst their goods were being dried, and till the women of the party were sufficiently recovered from cold and exposure to continue their journey to Drontheim, whither the whole party were bound; after which they left Thorfin, with many thanks for his courtesy and kindness. Grettir, · however, remained; not at the request of the bonder, who did not much like him, but to suit his own convenience. Indeed, he staid somewhat longer than Thorfin cared to keep him, considering what a fellow Grettir was, — never joining in conversation, unwilling to lend a helping hand in any work, a great stay-at-home, crouching over the fire all day, and, withal, eating voraciously. Thorfin was much out of doors: and, as he was a sociable man, he often requested Grettir to accompany him, either into the forest or about his farm; but could get no further answer than an impatient shake of the head and a grunt.

Now, the bonder was a fellow with a right merry heart and a kind one, and one, too, that loved seeing all around cheerful. With such a disposition, it is no wonder that the morose and indolent Grettir found no favor.

Yule drew nigh; and Thorfin busked him to depart, with a number of his freedmen, to keep high festival at one of his farms, distant a good day's journey. His wife was unable to accompany him, as the eldest daughter was ill, and wanted careful nursing; and Grettir was not invited, as his sullenness would have acted as a damper to the joviality of the banquet.

The farmer started for his farm in Slysfjord some days before Yule, accompanied by his thirty freedmen, expecting to meet a goodly throng of guests whom he had invited from all quarters.

Norway had for some time been in a disordered condition, from the mischief caused by numerous berserkirs and corsairs who roved over the country, challenging bonders to mortal combat for their homes, their wives and families. If a bonder declined to fight, as the law stood, his all was forfeited to the challenger; and if he fought, and was worsted, he lost his life as well. With the advice of Thorfin, Earl Erik Hakon's son put down these holm-bouts, and outlawed those whose custom it had been to make a business of them, going round the country and riding rough-shod over the peaceful bonders.

Among the worst of these were two brothers, well known for their wickedness, — Thorir wi' the Paunch, and Bad Ögmund. They were said to be stronger built than most, and to care for no man under the sun. They robbed wherever they went, burned farms over the heads of the sleeping inmates, and with the points of their spears drove the shrieking wretches back into the flames. When these pirates wrought themselves up into their berserkir rages, they howled like wolves, foamed at the mouth, their strength was increased to that of *trolls*,* and they rushed about, demon-possessed, murdering and destroying every living being that came in their way. Thorfin had been the prime instigator of their outlawry through the length and breadth of Norway; and, as may well be conjectured, the brothers bore him no good will, and only waited for an opportunity of wreaking their vengeance upon him.

The eve of Yule was bright and sunny; and the sick girl was so far recovered as to walk out and take the air, leaning on her mother's arm.

Grettir spent the whole day out of doors, in none of the sweetest of tempers at being excluded from the festivities of the season, and left to keep house with the women and eight dunder-headed churls. He fed his discontent by sitting on a headland, watching the boats glide past, as parties went to convivial gatherings at the houses of their friends. The deep-blue sea was speckled with white sails, as though countless gulls were playing on the waters. Now a stately dragon-ship rolled past, her fearful carved head glittering with gold and color, her sails spread like wings before the breeze, and her banks of oars flashing in the sun, then dipping into the sea. Now a wherry rowed by, laden with cakes and ale; and the boatmen's song rang merrily through the crisp air.

* Mountain-demons.

The day began to draw in; but still the red sparks from little vessels fleeting by in the dusk showed that all guests had not yet reached their destination.

Grettir was on the point of returning to the farm, when the strange proceedings of a craft at no great distance attracted his attention. He noticed that she stole along in the shadows of the islets, and darted with velocity across the open-water straits between them: she hugged the shore wherever she could, moved in a zig-zag course, and suddenly came flying with quick oar-sweeps towards the bay which Grettir was overlooking. In the twilight he could make out thus much of her, — that she floated low in the water, that she was built for speed, and that her sides were hung with shields. As she stranded, the rowers jumped on the beach. Grettir counted them, and found that they were twelve; armed men too! They broke into Thorfin's boat-house, and dragged forth his great punt, in which thirty men were wont to sit, pushed it out into deep water, and drew their own boat under cover, and pulled her up on the rollers.

Mischief was a-brewing: that was plain as a pike-staff! So Grettir descended the hill, and sauntered up to the band, with his hands in his pockets, kicking the pebbles before him, and humming a tune with the utmost nonchalance. "May I ask who is the leader of this party?" quoth he.

"Ah, ah! I'm the man," responded as ill-looking a fellow as Nature could well turn out of her laboratory. "Why, I am Thorir wi' the Paunch; and here's my brother Ögmund with all his rascals. I reckon the Bonder Thorfin knows our names; don't you think so, brother? And we have a little account to settle with him. Pray, is he at home?"

"Upon my word, you are lucky fellows," spoke Grettir, "coming here in the very nick of time, if you are the men I take you for. The bonder is from home with all his freedmen, and won't be back till after Yule : his wife and daughter, however, are at the farm. Now's your time, if you have old scores to wipe off; for there is every thing you can possibly want at the house, — silver, good clothes, ale, and provisions in the greatest profusion."

Thorir held his tongue whilst Grettir talked : afterwards he turned to his brother Ögmund, and said, "This is just what I expected; is it not? Now we can serve Thorfin out in thorough earnest for having made us outlaws. What a chatterbox this fellow is! There's no need of pumping to get any thing one wants to know out of him."

"Every man is master of his own tongue," retorted Grettir. "Now come along with me, and I will do the best I can for you."

The rovers thanked him, and accepted the invitation : so Grettir, taking Thorir by the hand, led him towards the farm, talking the whole way as hard as his tongue could wag. The housewife happened at the moment to be in the hall, putting up the hangings, and preparing for the Yule banquet; and, hearing Grettir speaking with such volubility, she stood still in astonishment, and asked whom he was greeting so cordially.

"It is quite the correct thing to receive guests well, is it not, mother?" asked Grettir : "and here are Thorir o' the Paunch, Bad Ögmund, and ten others, who have kindly come to join us in our Yule carousal; which is delightful; for, without them, our party would have been wofully scant."

"O Grettir! what have you done?" cried the poor woman. "You have brought hither the greatest ruf- fians in Norway. I would have given any thing that they had never come. This is the way in which you return the good Thorfin has done you in rescuing you from shipwreck, in taking you into his house, and car- ing for you through the winter as though you were one of his freedmen, and when you had not a farthing in your pocket to bless yourself withal!"

"Stop this abuse!" growled the young man. "There's time enough for that sort of thing another day. Now come and take off the wet clothes from the guests."

"You need not scream before you are hurt, my good woman," quoth Thorir: "you will want all your words for to-morrow, when I shall carry you and your daughter away with me, and you will have to say good-by to home for many a day. What think you of that?"

"Capital!" roared Grettir. "That is capital!"

On hearing this the housewife and her daughter fled to the women's apartment, crying, and wringing their hands with despair.

"Well," said Grettir, "as the women won't attend on you, I suppose that I must: so be good enough to hand me over any thing you want to have dried, such as your wet clothes and weapons."

"You're different from every one else in the house," spoke Thorir. "I almost think that you would make a boon companion."

"As you please," answered the young man. "Only, I tell you, I don't behave like this to all folk."

Then the freebooters gave him up their weapons: he wiped the salt water from them, and laid them aside in a warm, dry spot. Next he removed their wet garments, and brought them dry suits, which he routed out of the clothes-chests belonging to Thorfin and his freedmen.

By this time it was quite night. Grettir brought in logs, raked up the fire, and made a noble blaze.

"Now, my men," quoth he, "sit at table, and drink; for, i' faith, you must be thirsty after all the rowing you have done in the day."

"We are ready," said they: "only we don't know where to find the cellars."

"Will you let me fetch ale for you? or will you help yourselves?"

"Oh, go after it yourself, by all means!" answered they.

So Grettir brought the strongest ale, and poured out for them. The fellows were very tired, and drank copiously. Grettir stinted them neither in meat nor in drink; and at last he sat down at the end of the table and recited merry *sagas*, which riveted their attention, and delighted them amazingly. First he told the history of Hromund Greipsson, — how he broke open the tomb of the old Viking Thrain, and descended into it; how he wrestled with the demon-possessed corpse in its vault, and bore off its sword like a sunbeam; and how, in after-years, Hromund fought on the ice, and received fourteen wounds, lost his eight brothers, and, worst of all, saw his bright-flashing sword sink through an ice-floe. After that Grettir told the tale of An the Bow-brandisher, who would not turn his bow to enter the king's hall, but walked forward with it, though the horns stuck in the door-posts, and the bow bent nearly double, but did not break.

Not one of the house-churls showed his face in the hall that evening : they slunk about the farm, frightened and trembling.

Quoth Thorir, "I'll tell you what, comrades!—this lad is one of the best fellows I've clapped eyes on. I don't think we could meet in a hurry with another who would wait on us so well. What shall we give him?—Come, man, ask a boon of us!" Grettir answered, "I demand only one thing,—that, if we are as great allies in the morning as we seem to be to-night, I may become one of your gang. Even if I be weaker than the rest of you, be assured I will not hang back in the day of trial."

The pirates were delighted with this proposal, and wanted to clinch brotherhood at once; but Grettir objected. "No, no!" said he. "When liquor is in, wits leak out : you may come to a different mind in the morning, when you are sober, and regret what you have done. There is no need of hurry; and, as we are none of us famous for our discretion, a little thinking the matter over first is advisable."

They all protested that they would not change their opinion of him in the morning. Grettir, however, remained firm in his decision.

The young man saw now that they were getting rather tipsy: so he suggested that it was time for bed. "Yet first," said he, "you will, I know, like to run your eyes over Thorfin's storehouse."

"That we shall!" exclaimed Thorir, jumping up. "Come along, my lads!—follow me!"

Grettir took a lamp, and led the way.

The storehouse was separate from the house, and stood at right angles to it. It was a strongly-built place, made of large logs mortised firmly together: the door was also remarkably massive, and was furnished with a strong fastening. Adjoining this building was a lean-to office, divided off from the storehouse by a partition of planks: a flight of steps led to the office-door; for the house stood on a breast-high stone foundation.

The sharp, frosty air of night, striking on the faces of the revellers, increased their intoxication; and they became very disorderly, running against each other, uttering discordant whoops, and jolting Grettir's arm, so that he could with difficulty prevent the lamp from being knocked from his hand and extinguished.

Drawing back the bolt, he flung the door open, and showed the twelve men into the house. Then, slinging the lamp to a hook in one of the rafters, he let the rovers scramble for the prizes. The store was filled with various household goods, piles of costly garments, enamelled baldrics, carved and silver-mounted drinking-horns, some choice bracelets, and several bags, each containing a hundred ounces of pure silver. The drunken men were soon engaged in violent altercation over the spoil, as several coveted the same articles. In the midst of the hubbub Grettir stepped outside, closed the door, and bolted it. The freebooters did not notice his escape, as he had left the lamp burning; and they supposed that the door had swung to in the wind: they

10

were, moreover, too intent on selecting their shares of the booty to think of any thing else.

Grettir flew across the homestead to the farm-door, and cried loudly for the housewife; but she was silent, as she very naturally mistrusted his intentions, and had, besides, secreted herself, from fear of the pirates.

"Come, answer!" shouted Grettir. "I have captured the whole twelve; and all that is wanting is a supply of weapons! Call up the thralls, and arm them! Quick!—there is not a moment to be lost!"

"There are weapons enough here," answered the poor woman, emerging from her hiding-place. "But, Grettir, I have no faith in you."

"Faith or no faith," exclaimed Grettir, "I must have weapons at once! Where are the churls?—Here, Kolbein, Svein, Gamli, Krolf! Confound the rascals! where have they skulked to?"

"It will be a mercy of God if any thing can be done," said the housewife; "for we are in a sorry plight, to be sure! Now look here. Over Thorfin's bed hangs an enormous barbed spear: you will find there also helmet and cuirass, also a beautiful cutlass. No lack of weapons, if you have only the pluck to use them."

Grettir seized the casque and spear, girded on the sword, and dashed into the yard, begging the woman to send the churls after him. She called the eight men, and bade them arm at once, and follow. Four of their obeyed, rushing to the weapons, and scrambling for them; but the other four ran clean away.

I must tell you, that, in the mean time, the berserkirs

had rather wondered at Grettir's disappearance; and, from wondering, had fallen to suspecting that all was not right. Then they sprang to the door, tried it, and found it locked from without. It was too massive for them to break open; so they tore down the partition of boards between the store and the office. The berserkir rage came on them, and they ground their teeth, frothed at the mouth, and burst forth with the howl of demoniacs through the office-door, upon the landing at the head of the steps, just as Grettir came to the foot.

Thorir and Ögmund were together. In the fitful gleams of the moon they seemed like fiends, as they scrambled forth armed with splinters of deal, their eyes glaring with frenzy, and great foam-flakes bespattering their breasts, and dropping on the stones at their feet. The brothers plunged down the narrow stair with a yell which rang through the still, snow-clad forest for miles. Grettir planted the spear in the ground, and caught Thorir on its point. The sharp, double-edged blade, three feet in length, sliced into him, and came out beneath his shoulders, then tore into Ögmund's breast a span-deep. The yew-shaft bent like a bow, and flipped from the ground the stone, against which the butt had been planted. The wretched men crashed to the bottom of the stair, tried to rise, staggered, and fell again. Grettir planted his foot on them, and wrenched the blade from their wounds, drew the cutlass, and smote down another rover as he broke through the door. Other berserkirs poured out; and Grettir drove at them with spear, or hewed at them with sword: he slew another as the churls came up. They were late; for

they had been squabbling over the weapons: and, now
that they were come, they were nearly useless, as they
only made onslaughts when the backs of the robbers
were towards them; but, the moment that the vikings
turned on them, they bounded away, and skulked behind
the walls.

The pirates showed desperate fight, armed with chips
of plank or sticks pulled from some pine-fagots which
lay in the homestead. They warded off Grettir's blows,
and fled from corner to corner, pursued by their indefatigable foe. In the wildness and agony of despair
they could not find the gate, but bounded over the wall
of the yard, and ran toward the boat-house, with Grettir
at their heels. They plunged in, and possessed themselves of the oars : Grettir followed into the gloom, and
smote right and left. The bewildered wretches climbed
into the boat: some strove to push her into the water,
whilst others battled in the darkness with their unseen
enemy; but some pulled one way, some another, and
the blows from the oars fell on friend as well as foe, so
that the panic became more complete.

In the mean time the thralls had quietly returned to
the farm, quite satisfied when they saw the robbers take
to their heels; and no entreaties of the housewife could
induce them to follow Grettir. The four churls had had
quite enough of fighting. True, they had killed no one,
but then they had seen some men killed. Grettir sprang
into the boat, and stepped from bench to bench, driving
aft the terrified vikings. As the boat-house was open to
the air on the side which faced the sea, whilst the farther
end was closed with a door, Grettir was in shadow, whilst

the black figures of the rowers cut sharply against the moonlight, so that he could see where to strike, whilst his own body was undistinguishable.

One stroke from an oar reached him on the shoulder, and for the moment paralyzed his left arm. He killed two more vikings; and then the remaining four burst forth, and, separating into pairs, fled in different directions. Grettir followed the couple which was nearest, and tracked them to a neighboring farm, where they dashed into a granary, and hid among the straw. Unfortunately for them, most of the wheat had been threshed out, so that only a few bundles remained. Grettir shut and bolted the door behind him, then chased the poor wretches like rats from corner to corner till he had cut them both down; then he pulled the corpses to the door, and cast them outside.

In the mean while the sky had become overcast with a thick snow-fog which rolled up from the sea; so that Grettir, on coming out, saw that it would be hopeless attempting to pursue the two remaining berserkirs: besides, his arm pained him, his strength was failing him, and there stole over him an overpowering sense of weariness after his protracted exertions. The housewife had placed a lamp in the window of a loft; so that Grettir, seeing the light, was able to find his way back through the snow-storm without difficulty. When he came to the door, she met him, and, extending both her hands, gave him a cordial welcome. "You have indeed shown great valor," quoth she. "You have saved me and my household from insult and ruin. To you, and you alone, are we indebted."

"I am not much altered from what I was last evening; yet now you sing quite a different strain: then
you abused me most grossly," grumbled the young
man.

"Ah! but we little knew your mettle then. Come, be
a welcome guest within, and tarry till my husband re-
turns. Thanks are all that I can render you; but be assured Thorfin will not rest content till he has rewarded
this deed of yours munificently."

Grettir replied that he cared little for a reward, but
that he gladly availed himself of her invitation. "And
now I hope you may sleep without much fear of berserkirs." Grettir drank little, but lay down fully armed
for a sound and well-earned sleep.

On the following morning, as soon as day broke, a
party was formed to search for the two remaining vikings
who had escaped from Grettir in the darkness. The
snow had fallen so thickly during the night, that the
ground was covered, and all traces were obliterated; so
that the search proved ineffectual till dusk, when the
men were discovered under a rock, dead from cold and
loss of blood. The bodies were removed to the shore,
and buried under a cairn between tides.*   Then all re
turned to the farm in high glee; and Grettir chanted the
following verse: —

---

* Burial between tides was looked upon as disgrace: hence the Gula
Thing's law commands, "Every dead man is to be taken to church, and
buried in consecrated ground; except vile evil-doers, betrayers of their
masters, inveterate murderers, breakers of promised peace, thieves, and
suicides. Those men who have been guilty of the aforesaid crimes shall be
buried within reach of the tides, where the water licks the green turf."

" Twelve war-flame branches are buried
Low by the loud-resounding :
Unasked sent I them singly
To speedy death.   O ye gold-sallows,
Well born, bear me all witness !
What is wrought mightier?   Tell me
If ye wot, this being little." *

"There are not many men like you, certainly," answered the lady; "at all events, in this generation."
Then she seated him on the high stool of honor, and
treated him with every distinction.

So passed the time till the return of the bonder.

It was not till the Yule festivities were well over that
Thorfin busked him for return; then, after having dis-
missed his guests with presents, he and his freedmen
started for home, before news had reached him of what
had taken place during his absence.   The first startling
circumstance was the appearance of his great punt,
stranded.   Thorfin bade his men row to land with all
speed, as he suspected that this could not be the result
of accident.   The bonder was the first, in his anxiety,
to leap ashore and run to the boat-house.   There he saw
a ship hauled up on the rollers; and, at the second
glance, he knew it to be that of the vikings.   His cry
of dismay brought the rest around him.   He pointed to
the vessel, and said, "The red-rovers have made an at-

* I give this verse nearly literally, as a specimen of the curious style
of Icelandic poetry of the period.   *War-flame* is a periphrasis for a sword;
branch, or grove, for man: consequently, war-flame branch is a swords-
man.   *Gold-sallow* is similarly a periphrasis for woman; *loud-resounding*, for sea.

tack on my farm. I would give house and lands that
they had never come."

"What cause is there for fearing that a hostile visit
has been paid?" asked some of the men.

"I know whose boat this is," answered the farmer.
"It belongs to Thorir o' the Paunch and Bad Ögmund,
the two wickedest and most brutal of all the Norwegian
pirates. No effectual resistance can have been offered,
I fear, as the farm was deserted by all fighting-men, except, perhaps, that Icelander; but I put no trust in him
whatsoever."

The freedmen now consulted with the farmer as to
what steps should be taken, supposing that the house
were occupied by pirates.

All this while Grettir was at home, and he was to
blame for leaving Thorfin in uncertainty and alarm. He
had seen the master's boat round the headland, and enter the bay; but he would neither go himself to meet him
on the strand, nor suffer the thralls to do so.

"I do not care even though the bonder be a little dis-
tracted at what he sees," said the young man.

"Have you any objection to my going to the shore?"
asked the wife.

"None in the least: you are mistress of your own
actions."

Then she with her daughter ran to meet her husband,
and greeted him with a bright smile on her face. He
was delighted at seeing her, and said, kissing her forehead, "God be praised, sweetheart, that you and my
child are safe and sound! But tell me how matters have
stood during my absence; for, from the looks of affairs, I

do not think that you can have been left quite undisturbed."

"No more have we," she replied. "We have been in grievous danger of loss and dishonor; but the shipwrecked man whom you have sheltered has been our helper and guardian."

Thorfin said, "Sit by me on this rock, and tell me of what has taken place."

Then they took each other's hands, and sat together on a stone. The freedmen gathered round, and she told plainly and truthfully the story of the rovers, and Grettir's gallant conduct. When she spoke of the manner in which the young Icelander had decoyed them into the storehouse, and fastened them in, all the freedmen raised a shout of joy; and, when her tale was ended, their exultant cries rang so loud, that Grettir heard them in the farmhouse.

Thorfin spoke no word to interrupt the thread of his wife's recital; but the workings of his heart were clearly legible on his countenance. After she had ceased, he sat still, and rapt in thought: no one ventured to disturb him. Presently he looked up, and said, "The old saying proves to be true, 'Despair of no man.' Where is Grettir?"

"At home," answered the wife. "He is a strange man, and would not come to meet you."

"Then let me go to him," said the farmer, rising, and walking toward the house, followed by his men.

When he saw Grettir, he sprang to him, and thanked him in the fairest words for the heroism he had displayed.

"This I say to you," spoke Thorfin, "which few would say to their dearest friends, — that I hope one day you may need support, so as to prove how earnestly and joyfully I will strain every nerve to assist you; for assuredly I never can repay you for what you have done in my behalf till you are brought into great straits yourself. Abide with me as long as you list, and you shall be held in highest esteem by me and my followers."

Grettir thanked him heartily, and spent the rest of the winter at his house. The story of his exploit was noised throughout Norway, and it was especially praised on the spots where the berserkirs had given any trouble.

CHAPTER IX.

THE morning of the 2d of July was fair and sunny. We took an early breakfast. The farmer's son drove up our horses from the pasture, and we set off to visit the sulphur-jokuls about the Namarfjal. Down to the right lay the shining *Myvatn*, with its black islets; so many lava-ledges showing above the surface, about which scores of water-fowl were wheeling and clamoring. Ten minutes took us out beyond the farm-limits upon rising ground, down which the fiery floods had poured from the craters. If the reader has ever seen the heaps of slag, cinders, and clinkers, which come from the iron furnaces and mills, he can form a good idea of the sort of rock and soil which here covers the earth. We had to walk the horses, for the ground was full of holes, which were no less than bubbles in the lava, puffed up and burst at the top, leaving cavities often two and three feet in depth, with sharp, knife-like edges. Bad places for ponies' legs, these! Heaps of slag lay

155

piled all about, — red, brown, black, and vitreous Half
an hour over the ascending slope took us to the imme-
diate peaks.   Not a plant, shrub, nor a blade of grass, is
here to be seen, — nothing but bare volcanic rocks of an
almost pure red color.

"Look as if they were still red-hot, just from the furnace!" was Kit's impression as we sat gazing up at
them.

"I wonder why these sulphur-deposits are not worked
for commercial purposes," Raed queried.

Young Havisteen told us that there had been one attempt to do so, which had failed from some cause or
other.

"If managed properly, and worked economically, I
have no doubt there might be money made here," Kit
remarked.   "Havisteen, let's start some sulphur-works,
and make our fortunes."

But the young Dane didn't think he should like the
brimstone-business.   It was too suggestive, he said.   Besides, he didn't much like the place : it was altogether too
near a very fiery, restless region, situated at no great
depth below, where they had their brimstone hot instead
of cold.   He preferred, for permanent residence, a point
where the crust of the earth was thicker.

"That's because you're not a Yankee!" exclaimed
Wash.   "Why, down in our country, the folks take
worse risks of brimstone than this, every day, for not
half the prospect these jokuls offer : they do it for fun,
even!"

"Ah, yes! I've often heard that you are a very adven-
turous people," replied Jan.   "I hope you use needful

precautions, — such as paying the priest well, and getting your sins pardoned regularly."

A little farther on we entered a glen, or ravine, between two sharp ridges; and ere long, as we rode forward, a strange sound burst on our ears.

"What's that, Jan?" I asked.

"Sounds like a steamer-whistle," observed Raed.

"I think it is a steam-whistle of some kind," replied Jan.

On turning a bend in the hollow, we saw that the air had a misty, steamy look farther on. Presently we passed a small fissure in the crumbling tuff and bole, whence steam was emitted in quick, sharp puffs. All about the orifice was of a yellow hue.

"There's sulphur in that steam, you see," Jan remarked. "Sulphur is being deposited there."

A few rods farther on we passed two more of these *fumeroles*, as Raed termed them. Meanwhile the deep grum shriek of the whistle grew louder and more hideous, making the bare, rocky sides of the glen resound alarmingly; and about two hundred yards farther on we came in sight of a white jet of steam, streaming out from the face of a wild, black crag, rent and jagged with lavafragments. The sight of it gave us queer sensations. The steam spurts from a small crevice with vast impetuosity and a shuddering motion fearful to gaze upon. The wild, gruff shriek of the resonant edges utterly drowned our voices. The horses reared and plunged, snorting frightenedly. The ground all along the glen was warm; and judging from the color, the crumbling rocks, and the lava-clinkers, has clearly been in an intensely-heated condition some time.

Finding it quite impossible to get the horses past the whistle, we took them back to where a clump of low willow-shrubs were growing, fostered, no doubt, by the unnatural warmth of the soil. To the pliant twigs of these we twisted our bridles, and, leaving the scared nags, went back to prosecute our investigation on foot. Along the glen we found scores of the *fumeroles* puffing spitefully; and not only in the glen, but far up the sides of the mountains. The edges of the orifices are quite hot. Raed got his fingers scalded by getting them too far into the cracks of one of these. We stopped the mouths of several by way of experiment; but the steam soon made its way past the obstructions. In one case the stones were blown out with some violence. Kit and Raed climbed up the crag to the whistle. They reported, that, at a distance of three or four yards, the noise was perfectly deafening; and also that the rocks were too hot for their hands within a radius of six feet of the hole. There must be a terrific head of steam on somewhere in behind that crag.

At another place, during our ride along these sulphur-ridges, we came upon a block of pure brimstone, from which we broke off bits as specimens for our collection of minerals. The soil all about this place was of soft bole of a brick-red color. So many objects of curiosity arrested us all along our way, that it was four o'clock (afternoon) ere we came to the Namarskarth, or *pass*, which leads into the famous plain beyond the sulphur-ridges. Rather reluctantly, therefore, we gave up visiting it till the next day. We had taken nothing save a lunch with us, and now began to long exceedingly for the gooa cheer which we knew would be awaiting us at the farmer's *byre*.

After supper, and during the long glowing evening, the farmer read to us another chapter from the "Story of Grettir;" which I subjoin without further preface.

Grettir in Norway.

There was a man named Thorir, who lived at Garth in Athaldal. He was a mighty Icelandic chief, with numerous retainers and extended influence. He had two sons, —fine promising fellows both of them, and pretty nearly full-grown men. Thorir had spent the summer in Norway when King Olaf returned from England, and had got into favor with the king, and also with Bishop Sigurth, as may be judged by the fact, that Thorir, after having built a ship, asked him to consecrate it; which was a great condescension on the part of Thorir.

Thorir left Norway for Iceland: he reached it safely, and then chopped up his boat, as he was tired of the sea. The two beaks of the prow he set up over his hall-doors: and they were sure indications of the direction of the wind; for the north wind piped in one, and in the other wailed the south wind.

As soon as the news reached Iceland that King Olaf was supreme over the whole of Norway, Thorir considered that there might be a good opening at court for his two sons: so he packed them both off late in the autumn to pay their respects to the king, and remind him of his old friendship for their father.

They landed in the south of Norway; and then, get

ting a long rowing-boat, they skirted the coast on their
way north to Drontheim.   Reaching a fine frith in
which there was shelter from the gales, which began to
bluster violently as the winter drew nigh, the sons of
Thorir ran their boat in, and determined on waiting till
the storms blew over in a comfortable hostel, built some
way up the shore for the accommodation of travellers.
Their days they spent in hunting bears among the
mountains, and their nights in merry carousal.

It happened that Grettir was on board a merchant-
man, then off the shores of Norway, beating about in
the gale, seeking safe harborage.

Late one evening the vessel ran up this same fiord,
and stranded on the side opposite that on which was the
hostel.   The night was cold and wintry : heavy storms
of snow rolled over the country, whitening the mountains, and forming drifts behind the rocks.   The men
from the ship were worn out, and numbed with cold ;
and they knew not on what part of the coast they had
stranded.

When they reached land, they hurried from the shore
to seek a sheltered nook where they might pass the
night.

It was a wild night.   The moon had been clouded
over by piles of gray mist, which rolled through the sky,
sending out arms of vapor.   Haggard and ghastly, she
seemed to steal over her course swathed in grave-clothes.
Now and then some crags caught a straggling gleam,
and flashed forth, but, directly after, were again blotted
out; then the fiord caught the light, and shone like
steel till the shadows turned it to lead.   An uncertain

light flickered down the mountain-side over the pine-forests, which raved and bent as the wind poured through them.

Suddenly a spark, then a flame, was distinguishable, twinkling among the trees on the opposite side of the fiord. This was a tantalizing sight for the poor shivering fellows; and they began to wish that some one of their number would swim across, and bring over a light. No one, however, offered; and the crew hesitated about pushing the ship off and rowing across, lest they should fall among rocks, and injure the vessel.

"In the good old times, there must have been men who would have thought nothing of swimming across the frith by night," said Grettir.

"Maybe," answered some of the party; "but it is of no odds to us what men have been, if there are none now up to the mark. Why do you not venture yourself, Grettir? You are as strong and plucky as any of the old heroes. You see what straits we are put to for want of a little fire."

"There is no great difficulty in procuring a light," answered the young Icelander; "but I know that I shall get no thanks for my pains."

"Then you must have an uncommonly poor opinion of us," said the chapmen.

"Well," quoth Grettir, "I will risk it: at the same time, I tell you I have a presentiment that you will bear me no good will for what I do."

They pooh-poohed his objections, and assured him that he was the best fellow going.

Then Grettir flung his clothes off, and busked him for

11

swimming. He had on him a fur cape, and a pair of *wadmal* breeches : these he hitched up, and strapped tightly round his waist with a bark cord; then, catching up an iron pot, he jumped into the sea, and swam across. On reaching the farther side, he stood up on the beach, and shook the superfluous water from him; but before long his trousers froze hard, and the water formed in icicles round the hood of his cape.

Grettir ascended through the pine-wood towards the light; and, on reaching the hostel from whence it proceeded, he walked straight in without speaking to any one, and, striding up to the fire, stooped, and began to rake the embers into his iron pot, and to select a blazing brand which he could carry across in his mouth. The hall was full of revellers, and these revellers were the sons of Thorir and their boat's crew. They were already half intoxicated; and on seeing a tall, wild-looking man enter the hall, half dressed in fur, and bristling with icicles, they concluded at once that they saw a *troll*, or mountain-demon.

Whereupon every man caught up the first weapon he could lay hold of, and rushed to the attack. Grettir defended himself as best he could, warding off the blows with the flaming log, and eluding the missiles flung at him. In the scuffle, the hot embers on the hearth were scattered over the floor, which was strewn with fresh straw and rushes.

In a few moments the hall was filled with flame and smoke; and Grettir broke through it, escaped to the shore, plunged into the waves, and reached the other side in safety.

He found his companions waiting for him behind a rock, with a pile of dry wood which they had collected during his absence. The cinders were blown upon, and twigs applied, till a blaze was produced; and before long the whole party sat rubbing their almost-frozen hands over a cheerful fire.

On the following morning the merchants recognized the fiord; and, remembering that on its bank stood the house of refuge which King Olaf had built for weather-bound travellers, they supposed that the light Grettir had procured must have come from it: so they determined on running the boat across, and seeing who were then quartered in the hostel.

When they reached the spot, they found nothing but an immense heap of smoking ashes. From under some of the charred timber projected scorched human limbs. The chapmen, in alarm and horror, turned upon Grettir, and charged him with having maliciously burned the house with all its inmates.

"There now!" exclaimed Grettir. "I had a presentiment that misfortune would attend my undertaking last night. I wish that I had not taken so much trouble for a set of thankless churls like you."

The ship's crew raked the embers out, and pulled aside the smoking beams in their search for the bodies, that they might give them decent burial. In so doing, they came upon some whose features were not completely obliterated; and among these was one of the sons of Thorir. It was at once concluded that the party brought by Grettir to such an untimely end was that of Thorir's sons, which had sailed shortly before the chapmen. The indignation of the merchants became so vehement, that they drove Grettir with imprecations from their company, and refused to receive him into their vessel for the remainder of the voyage.

Grettir, in sullen wrath, would say no word in self-defence; but, turning on his heel, he stalked proudly into the woods, with his sword by his side and his battle-axe over his shoulder, determined on exculpating himself before King Olaf, and him alone. The vessel reached Drontheim before him; and the news of the hostel-burning caused universal indignation.

One day, as the king sat at audience in his hall, Grettir strode in, and, going before Olaf, greeted him. The king eyed him all over, and said, —

"Are you Grettir the Strong?"

He answered, "Such is my name; and I have come hither, sire, to get a fair hearing, and rid myself of the charge of having burned men maliciously. Of that I am guiltless."

Olaf replied, "I sincerely hope that what you say is true, and that you will have the good fortune to clear yourself of the imputation laid against you."

Grettir said that he was willing to do any thing the king wished in order to prove his innocence.

"Tell me, first," quoth the king, "what is the true version of the story, that I may know what steps are to be taken."

Grettir answered by relating all the circumstances, and he asserted that the men were alive when he left the hostel, carrying the fire.

The king remained silent for some moments.

"If I might fight some one!" suggested Grettir: "I should rather like it."

"I have no doubt that you would," replied Olaf. "But, remember, you have not a single accuser, but a whole ship's crew; and you cannot fight them all."

"Why not?" asked the Icelander: "the more, the merrier. Let them come!"

"No, no, Grettir!" answered the king. "I cannot allow such a proceeding to take place. But I will tell you what you shall do, — go through the fire-ordeal."

"What is that?" asked the young man.

"You must lift bars of iron heated till the furnace can make them no hotter, and walk with bare feet on red-hot ploughshares."

"I'll do it at once!" said Grettir. "Where are the ploughshares?"

"Stop!" said the king. "You would be burned, to a certainty, if you ventured without preparation."

"What preparation?" asked Grettir.

"A week of prayer and fasting," was the reply.

"I do not like fasting," said the young man.

"But you cannot help yourself," answered Olaf.

"I cannot pray," said Grettir: "I never could."

"Then the bishop shall teach you," answered the king, with a smile at the bluntness of the Icelander.

Grettir was removed, and kept in custody by the clergy, who did their best to prepare him for the solemn moment of the ordeal; but they found him a troublesome fellow to manage.

The day came; and Drontheim was thronged with people, who streamed in from all the country round to see the Icelander of whom such stories were told. A procession was formed. The king's body-guard marched at the head, followed by the king himself, the bishop, the choir, and the clergy; amongst whom walked Grettir, a head taller than any of the throng, upright, his wild brown hair flying loose in the breeze, his arms folded, and his honest blue eyes wandering over the sea of heads which filled the square before the cathedral-doors. The crowd pressed in closer and closer, but without in the slightest degree disconcerting him. Opinions seemed to be divided as to whether he were guilty or not: his dauntless bearing, and open, sunny countenance, were not those of a truculent berserkir. Among the mob was a young man of dark complexion, who made a great noise, wrangling, and shouldering his way till he reached the procession.

"Look at him!" exclaimed he. "This is the man, who, in cold blood, could burn a house down over helpless victims, and exult at their shrieks of despair; yet now is about to be given a chance of escape, when every one knows that he is a deep-dyed villain!"

"But he says that he is guiltless," quoth a man in the crowd.

"Innocent!" exclaimed the youth. "A plea of innocence has been set up as an excuse, because the king wishes to have him in his body-guard."

"He should have a chance of clearing his character," spoke a person standing near.

"Ay; but who knows how the irons may be tampered with by the king and clergy, so that this ruthless murderer may escape the punishment he deserves?"

"Young man," spoke Grettir with a voice like thunder, whilst flame leaped up in his eyes, and his strong
limbs quivered with rage, — "young man, beware!"

"Beware of what, pray?" laughed the youth.
"Though you may escape the punishment you have so
richly deserved, yet you shall not escape me."

And, springing up, he thrust his nails into Grettir's
face, so that he brought blood; calling him, at the same
time, son of a sea-devil, troll, and other insulting
names. This was more than the Icelander could bear:
he caught the young man up, shook him as a cat shakes
a mouse, and flung him to the ground with such
violence, that he lay senseless, and was carried away as
if dead.

This act gave rise to a general uproar. The mob
wanted to lay hands on Grettir; some threw stones;
others assaulted him with sticks: but he, planting his
back against the church-wall, rolled up his sleeves, and
guarded off the blows, shouting joyously to his assail-
ants to come on.

A flush of honest joy at the prospect of a fight
mantled in his cheeks, and his eyes sparkled with delight. Not a man came within his reach but was sent
reeling back, or felled to the ground.

Grettir caught a stick aimed at him while it was in
the air, and dealt such blows with it, that he cleared a
ring about him; whilst still, with a voice clear as a bell,
he called to the mob to come on manfully, and not
shrink back like cowards.

In the mean time the king and bishop had been
waiting in church; the processional psalm was ended;

the red-hot ploughshares were laid in the choir, and were gradually cooling: yet no Grettir came.

At the same time sounds of uproar entered the church, and the king sent out to know what was the matter. His messenger returned a moment after with a report, that, without the cathedral, the Icelander was fighting the whole town.

The king thereupon sprang from his throne, hastened down the nave, and came out of the great western door whilst the conflict was at its height.

"O sire!" exclaimed Grettir, "see how I can fight the rascals!" And, at the word, he knocked a man over at the king's feet.

"Hold, hold!" exclaimed Olaf. "What have you done, throwing away the chance of exculpating yourself from the charge laid against you?"

"I am ready now, sire," answered Grettir, wiping the perspiration and blood from his face, and smoothing down his hair, which was standing on end. "Let us go into the church at once: I am longing for the red-hot ploughshares."

He would have pushed past the king had not Olaf prevented him, saying that his opportunity was past, as he was guilty of mortal sin in having killed the young man who had assaulted him, and maimed so many other persons.

"What is to be done?" exclaimed Grettir. "I have undergone all that week of fasting for nothing. Sire, might not I become your hench-man? You will find me stronger than most men."

"True enough," answered the king. "Few men have

the strength and courage which you possess; but ill luck attends on you. Besides, I dare not keep you by me, as you would continually be getting into hot water. Now, this I decree : you shall be in peace during the winter; but with the return of summer you shall be outlawed, and go to Iceland, where, I forewarn you, you shall lay your bones."

Grettir answered, " I should like first to get rid of the charge of the hostel-burning; for, honor bright! I never intended to do the mischief."

"That is likely enough," said the king; "but it is quite impossible now for you to go through the ordeal."

After this, Grettir hung about the town for some while; but Olaf paid no further attention to him : so at last he went off to stay the rest of the winter with a kinsman.

On the return of spring, the news of what Grettir had done reached Iceland ; and, when they came to the ears of Thorir of Garth, he rode with all his friends and clients to Thing, and brought an action against Grettir for the burning of his sons. Some men thought that the action was illegal, as the defendant was not present to take exception. However, the end of the action was, that Grettir was outlawed through the length and breadth of Iceland. Thorir set a price on his head, and proved the bitterest of Grettir's foes.

Towards the close of the summer, Grettir arrived in a vessel off the mouth of the White River, in Borgar Fiord.

It was a still, summer night when the ship dropped anchor. The Skarths-heithi chain was purple; but

Baula's sharp cone was steeped in gold, and the distant silver cap of Ok shone in the sun's rays like a rising moon. The steam rising from the numerous springs in Reykholts-dall was rounded and white in the cool, still air. Flights of swans sailed overhead with their harp-like melody. As the gulls dipped in the calm water, every feather of their white wings was reflected. A boat came from shore, and was rowed to the ship.

Grettir stood watching it from the bows, leaning on his sword. As the smack touched the side of the ship, "What news?" he called.

"Are you Grettir, Asmund's son?" asked a man, rising in the boat.

"I am," replied Grettir.

"Then we bear you ill news: your father is dead."

Another man stood up in the boat, and said, " Grettir, your brother has been murdered."

" And you," called a third boatman, "have been out-lawed through the length and breadth of Iceland."

It is said that Grettir did not change color, nor did a muscle in his whole body quiver; but he lifted up his voice, and sang, —

> " All at once are showered
> Round me, rhyme collector,
> Tidings sad, — my exile,
> Father's loss, and brother's,
> Branching boughs of battle !
> Many blue-blade breakers
> Shall bewail my sorrow."

One night Grettir swam ashore, obtained a horse. and

reached the middle frith in two days. He arrived at home by night, when all were asleep: so, instead of disturbing the household, he went round to the back of the house, opened a private door, stepped into the hall, stole up to his mother's bed, and threw his arms round her neck.

She started up, and asked who was there. When he told her, she clasped him to her heart, and laid her head sobbing on his breast, saying, "O my boy! I am bereaved of my children! Atli, my eldest, is murdered; and you are outlawed: only my baby Illugi remains!"

Grettir remained at home for some days, till Thorir of Garth learned where he was, and then he was compelled to fly. He was hunted from place to place; and, to the last, Thorir remained his implacable enemy.

CHAPTER X.

ON the morning of the 3d we rode off to the Namarskarth at eight o'clock, and reached the entrance to the gorge at about ten. Passing through, we ere long came out to one of the strangest and most appalling spectacles which we had thus far witnessed. Vast volumes of steam were rolling up, and drifting in mighty whorls high over the enclosing jokuls. Amid these we got glimpses of a dark mud-plain, walled in by lava-fields. The steam-clouds seemed to gush mainly from this mud-flat; though the whole basin, up to the very tops of the mountains, was steaming steadily, depositing sulphur. The rocks about the mud, up to the very mountain-tops, were of a frightful primrose tint; and, seen through the whirling vapors, seemed like embers in a fiery grate. A curious brightness hovered over the landscape, which the shadowy volumes sweeping athwart the sun seemed scarcely to darken. Beneath this volcanic panorama, a low, deep, drumming sound, accompanied by an ominous jar of the earth, tells of the furious agencies at work. If any of our modern poets should find it desirable to portray the entrance to the *bad place*, I would respectfully recommend that they assist imagination by a visit to the Namarskarth. I am quite sure, that, well described, it would go far ahead, in point of hellishness, of any thing to be found in Dante. There is a certain diabolical aspect about this locality I have never yet seen surpassed. A strong odor of sulphur, and several other gases, moderately assist the fancy. Kit remarked, that if he had been out walking of a fine afternoon alone, and had come out upon this plain, he should have immediately made tracks for a less suspicious quarter of the country, without stopping to conduct an investigation as to natural causes.

After a great *to-do*, we at length got our horses as far as a clump of willows, which seemed strangely at home in this rather unpromising region. Here we made them fast, and left them in a snorting fright, while we pushed on out upon the plain. Quite a hedge of crane's-bill and other weeds has sprung up on the warm borders of this steamy tract. These cease, however, at a short distance out; and, on laying the back of my hand down to the soft clay, I found that it was much too hot for comfort. Indeed, a very genial warmth kept rising in our faces; and our feet were soon "hot as pepper." Not many days after, we made the discovery that the boots worn on this occasion would tear like brown paper. At the surface the clay seemed tolerably hard, and even tough; but it sank alarmingly under us in many places, especially when we stood near together. Going on, we began to get among a legion of little slobbering mud-pools, swelling up in bell-shaped bubbles, out of which puffed whiffs of steam and stinking gases. About all these were filmy rings of slime of ghastly bluish, saffron, and even greenish tints. Presently Raed, who had ventured on ahead, called to us. We stole cautiously forward. He was standing on the edge of a big caldron of boiling, spattering mud and slime, which stewed and simmered and hissed and groaned in a manner altogether horrible, — hellish. The caldron itself was, for a guess, twenty feet in diameter, and sunk four or five feet below the level of the surrounding plain. Portions of its banks were continuously crumbling. Even as we stood looking over, a part of the earth under our feet caved in. Kit, who stood nearest, came near going with it; and was only saved by a smart spring and our rescuing hands. After that, we judiciously kept back from the brink. As we ventured farther on among the bewildering steam-clouds, a noise as of an earthquake-rumble began to be heard from beneath the plain. Spurts of steam were streaming up all about. Another large caldron was strumming and humming in a constant tremble, with jets of ink-black water spouting up to the height of six or eight feet, before which the loose clay-banks crumbled, and melted like lard. As we stood watching this, a shrill steam-whistle burst out on a sudden. Kit at once started toward the sound; and the rest of us followed, stepping very cautiously among the seething puddles and scalding steam-jets. Ere long we came out upon the verge of another caldron of what looked to be very black molasses, boiling with a queer *plop-plop, plop-plop ;* while every few seconds a dome-shaped mud-bubble would burst with a spiteful *thut*, sending the hot spatters in all directions. On the farther edge of this slime-boiler, at a point where a shoot of small stones and pumice had rattled down the bank, a stiff little *spurt* of steam was darting out with amazing force : this was the whistle.

Turning to the right, we wended our way slowly back out of this steamy labyrinth, and, after passing numerous other caldrons and slime-pools, at length found ourselves at no great distance from where we had come on.

After a lunch, Wash and Kit went on again to make a sketch of the caldron, where the whistle was still heard screaming.

Raed and I climbed up among the rocks to get a better view of the strange landscape. What impressed me most, after the eye had grown a little accustomed to the curious scene, was the strange, unnatural medley of sounds, which, commingled, and utterly incomparable, came wafted up with the steam-clouds.

We set out on our return at a little after four, and arrived at Reykjalith at about six.

We were agreed in rating the scene at the Namarskarth more impressive than the geysers. Taken together, it leaves a far more vivid impression on the fancy.

The farmer was hoarse and *roopy* from a cold that evening: he could not read to us. We were not a little disappointed; for we had come to grow interested in the sturdy Grettir, whom the farmer plainly loved as his own son.

"I wonder whether the cherry-cheeked lass can read?" Raed queried.

Havisteen thought it likely. "All the children are taught to read at home," he said.

"Let's ask her to read to us," Kit proposed.

Young Havisteen thought this a rather hard question for him to ask; but finally went out to the kitchen. We waited the result with some curiosity. After a long while, he came back accompanied by the son of the farmer. Indridi — that was the girl's name — could not, by any manner of means, be prevailed upon to favor us with a reading; but Biarni had consented. We felt a little patronizingly toward Biarni, he seemed so much embarrassed; but he read well, clearly and fluently. The farmer was too indisposed to come in. Indridi and her mother joined us after a while. Our reading and note-taking went on swimmingly for some hours, disturbed a little at one time by the shameless Wash, who, not having the fear of the old farmer before his eyes this evening, waxed noisy in his odious attentions to the rosy Indridi. But he waked a reproof from an unexpected source. The old lady did not like to have a saga-reading thus disturbed. She turned upon them suddenly with a sharp word and a still sharper look. After that they were whist enough. Didn't that gratify Kit!

This evening Biarni read about

GRETTIR AND THE VAMPIRE.

In the beginning of the eleventh century, there stood, a little way up this valley (the Vale of Shadows),

a small farm, occupied by a worthy bonder named Thorhall and his wife. The farmer was not exactly a chieftain; but he was well enough connected to be considered
respectable. To back up his gentility, he possessed numerous flocks of sheep, and a goodly drove of oxen.
Thorhall would have been a happy man but for one
circumstance, — his sheep-walks were haunted.

Not a herdsman would remain with him. He bribed,
threatened, entreated, all to no purpose. One shepherd
after another left his service; and things came to such
a pass, that he determined on asking advice at the next
annual council. Thorhall saddled his horses, adjusted
his packs, provided himself with hobbles, cracked his
long Icelandic whip, and cantered along this identical
road; and, in less time than we have taken over it, he
reached Thingvellir.

Skapti, Thorodd's son, was lawgiver at that time; and
as every one considered him a man of the utmost prudence, and able to give the best advice, our friend from
the Vale of Shadows made straight for his booth.

"An awkward predicament, certainly, to have large
droves of sheep, and no one to look after them," said
Skapti, nibbling the nail of his thumb, and shaking his
wise head, — a head as stuffed with law as a ptarmigan's
crop is stuffed with blæberries. "Now, I'll tell you
what: as you have asked my advice, I will help you to a
shepherd; a character in his way; a man of dull intel-
lect, to be sure, but strong as a bull."

I do not care about his wits, so long as he can look
after sheep," answered Thorhall.

"You may rely on his being able to do that," said

Skapti. "He is a stout, plucky fellow; a Swede from Sylgsdale, if you know where that is."

Towards the break-up of the council, — "Thing" they call it in Iceland, — two grayish-white horses belonging to Thorhall slipped their hobbles, and strayed: so the goodman had to hunt after them himself, which shows how short of servants he was. He crossed Sletha-ási; thence he bent his way to Armanns-fell; and, just by the Priests'-wood, he met a strange-looking man, driving before him a horse laden with fagots. The fellow was tall and stalwart. His face involuntarily attracted Thorhall's attention: for the eyes, of an ashen gray, were large and staring; the powerful jaw was furnished with very white protruding teeth; and around the low forehead hung bunches of coarse, wolf-gray hair.

"Pray, what is your name, my man?" asked the farmer, pulling up.

"Glámr, an please you," replied the wood-cutter.

Thorhall stared; then, with a preliminary cough, he asked how Glámr liked fagot-picking.

"Not much," was the answer: "I prefer shepherd life."

"Will you come with me?" asked Thorhall. "Skapti has handed you over to me, and I want a shepherd this winter uncommonly."

"If I serve you, it is on the understanding that I come or go as pleases me. I tell you, I'm a bit truculent if things do not go just to my thinking."

"I shall not object to this," answered the bonder. "So I may count on your services?"

"Wait a moment. You have not told me whether there be any drawback."

' I must acknowledge that there is one," said Thor-hall : "in fact, the sheep-walks have got a bad name for bogies."

"Pshaw! I'm not the man to be scared at shadows," laughed Glámr: "so here's my hand to it. I'll be with you at the beginning of the winter night."

Well, after this they parted; and presently the farm-er found his ponies. Having thanked Skapti for his advice and assistance, he got his horses together, and trotted home.

Summer, and then autumn, passed; but not a word. about the new shepherd reached the Valley of Shadows. The winter-storms began to bluster up the glen, driving the flying snow-flakes, and massing them in white drifts at every winding of the vale. Ice formed in the shallows of the river; and the streams, which in summer trickled down the ribbed scarps, were now transmuted into icicles.

One gusty night, a violent blow at the door startled all in the farm : in another moment, Glámr, tall as a troll, stood in the hall, glowering out of his wild eyes, his gray hair matted with frost, his teeth rattling and snapping with cold, his face blood-red in the glare of the fire which smoldered in the centre of the hall.

Thorhall jumped up and greeted him warmly ; but the housewife was too frightened to be very cordial.

Weeks passed, and the new shepherd was daily on the moors with his flock. His loud and deep-toned voice was often borne down on the blast as he shouted to the sheep, driving them into the fold. His presence always produced gloom ; and, if he spoke, it sent a thrill through the women, who openly proclaimed their aversion for him.

There was a church near the *byre ;* but Glámr never crossed the threshold : he hated psalmody, which shows what a bad man he was.

On the vigil of the Nativity Glámr rose early, and shouted for meat. "Meat !" exclaimed the housewife : "no man calling himself a Christian touches flesh to-day. To-morrow is the holy Christmas Day, and this is a·fast."

"All superstition !" roared Glámr. "As far as I can see, men are no better now than they were in the bonny heathen time. Now bring me meat, and make no more ado about it."

"You may be quite certain," protested the goodwife, "if church rule be not kept, ill luck will follow."

Glámr ground his teeth, and clinched his hands. "Meat ! I will have meat, or " — In fear and trembling the poor woman obeyed.

The day was raw and windy : masses of gray vapor rolled up from the Arctic Ocean, and hung in piles about the mountain-tops. Now and then a scud of frozen fog, composed of minute spiculæ of ice, swept along the glen, covering bar and beam with feathery hoar-frost. As the day declined, snow began to fall in large flakes like the down of the eider-duck. One moment there was a lull in the wind; and then the deep-toned shout of Glámr, high up the moor-slopes, was heard distinctly by the congregation assembling for the first vespers of Christmas Day. Darkness came on, deep as that in the rayless abysses of Surtshellir ; and still the snow fell thicker. The lights from the church-windows sent a yellow haze far out into the night, and every flake burned golden as it swept within the ray. The bell in the lich-gate clanged for even-song, and the wind puffed the sound far up the glen: perhaps it reached the herdsman's ears. Hark! some one caught a distant shout or shriek: which it was he could not tell; for the wind muttered and mumbled about the church-eaves, and then, with a fierce whistle, scudded over the graveyard fence.

Glámr had not returned when the service was over. Thorhall suggested a search; but no man would accompany him. And no wonder: it was not a night for a dog to be out in; besides, the tracks were a foot deep in snow. The family sat up all night, waiting, listening, trembling; but no Glámr came home.

Dawn broke at last, wan and blear in the south. The clouds hung down like great sheets, full of snow, almost to bursting.

A party was soon formed to search for the missing man. A sharp scramble brought them to high land; and the ridge between the two rivers which join in Vatnsdalr was thoroughly examined. Here and there were found the scattered sheep, shuddering under an icicled rock, or half buried in a snow-drift. No trace yet of the keeper. A dead ewe lay at the bottom of a crag: it had staggered over it in the gloom, and had been dashed to pieces.

Presently the whole party were called together about a trampled spot in the heithi, where evidently a death-struggle had taken place; for earth and stone were tossed about, and the snow was blotched with large splashes of blood. A gory track led up the mountain, and the farm-servants were following it, when a cry, almost of agony, from one of the lads, made them turn

In looking behind a rock, the boy had come upon the corpse of the shepherd: it was livid, and swollen to the size of a bullock. It lay on its back, with the arms extended. The snow had been scrabbled up by the puffed hands in the death-agony; and the staring, glassy eyes gazed out of the ashen-gray, upturned face, into the vaporous canopy overhead. From the purple lips lolled the tongue, which in the last throes had been bitten through by the horrid white fangs; and a discolored stream which had flowed from it was now an icicle.

With trouble the dead man was raised on a litter, and carried to a gill-edge ; but beyond this he could not be borne. His weight waxed more and more. The bearers toiled beneath their burden; their foreheads became beaded with sweat : though strong men, they were crushed to the ground. Consequently, the corpse was left at the ravine-head, and the men returned to the farm. Next day their efforts to lift Glámr's bloated carcass, and remove it to consecrated ground, were unavailing. On the third day a priest accompanied them ; but the body was nowhere to be found. Another expedition, without the priest, was made ; and on this occasion the corpse was found : so a cairn was raised over it on the spot.

Two nights after this, one of the thralls, who had gone after the cows, burst into the *stofa* with a face blank and scared. He staggered to a seat, and fainted On recovering his senses, in a broken voice he assured all who crowded about him that he had seen Glámr walking past him as he left the door of the stable. On the following evening a house-boy was found in a fit

under the *tún* wall, and he remained an idiot to his dying-day. Some of the women next saw a face, which,
though blown out and discolored, they recognized as that
of Glámr, looking in upon them through a window of
the dairy. In the twilight Thorhall himself met the
dead man, who stood and glowered at him, but made no
attempt to injure his master. The haunting did not
end there. Nightly a heavy tread was heard around
the house, and a hand feeling along the walls, sometimes
thrust in at the windows; at others clutching at the
wood-work, and breaking it to splinters. However,
when the spring came round, the disturbances lessened,
and, as the sun obtained full power, ceased altogether.

That summer a vessel from Norway dropped anchor
in Hunavater. Thorhall visited it, and found on board
a man named Thorgaut, who was in search of work.

"What do you say to being my shepherd?" asked the
bonder.

"I should much like the office," answered Thorgaut.
"I am as strong as two ordinary men, and a handy fellow to boot."

"I will not engage you without forewarning you of
the terrible things you may have to encounter during
the winter night."

"Pray, what may they be?"

"Ghosts and hobgoblins," answered the farmer: "a
fine dance they lead me, I can promise you."

"I fear them not," answered Thorgaut. "I shall be
with you at cattle-slaughtering time."

At the appointed season the man came, and soon
established himself as a favorite in the household. He

romped with the children, chucked the maidens under the chin, helped his fellow-servants, did odd jobs for his master, gratified the housewife by admiring her *skyer*, and was just as much liked as his predecessor had been detested. He was a devil-may-care fellow too, and made no bones of his contempt for the ghost, expressing hopes of meeting him face to face; which made his master look grave, and his mistress shudderingly cross herself. As the winter came on, strange sights and sounds began to alarm the folk. But these never frightened Thorgaut: he slept too soundly at night to hear the tread of feet about the door, and was too short-sighted to catch glimpses of a grizzly monster striding up and down in the twilight before its cairn.

At last Christmas Eve came round, and Thorgaut went out as usual with his sheep.

"Have a care, man!" urged the bonder: "go not near to the gill-head, where Glámr lies."

"Tut, tut! fear not for me. I shall be back by vespers."

"God grant it!" sighed the housewife; "but 'tis a whist day, to be sure!"

Twilight came on. A feeble light hung over the south. Far off in southern lands it was still day; but here the darkness gathered in apace, and men came from Vatnsdalr for even-song to herald in the night when Christ was born. Christmas Eve! How different in Saxon England! There the great ashen fagot is rolled along the hall with torch and taper; the mummers dance with their merry jingling bells; the boar's-head with gilded tusks, "bedecked with holly and rosemary," is brought in by the steward to a flourish of trumpets.

How different, too, where the Varanger cluster round
the imperial throne in the mighty church of the Eternal
Wisdom at this very hour! Outside, the air is soft from
breathing over the Bosphorus, which flashes tremulously
beneath the stars. The orange and laurel leaves in the
palace-gardens are still exhaling fragrance in the hush
of the Christmas night.

But it is different here. The wind is piercing as a
two-edged sword. Blocks of ice clash and grind along
the coast of the Hunaflói, and the lake-waters are
congealed to stone. Aloft the aurora flames crimson,
flinging long streamers to the zenith, and then suddenly
dissolving into a sea of pale-green light. The natives
are waiting around the church-door; but no Thorgaut
has returned.

They find him next morning lying across Glámr's
cairn, with his spine, his leg and arm bones, shattered.
He is conveyed to the churchyard, and a cross is set up
at his head. He sleeps till the resurrection peacefully.

Not so Glámr: he becomes more furious than ever.
No one will remain with Thorhall now except an old
cowherd who has always served the family, and who had
long ago dandled his present master on his knee.

"All the cattle will be lost if I leave," said the carle:
"it shall never be told of me that I deserted Thorhall
from fear of a spectre."

Matters rapidly grew worse. Outbuildings were
broken into of a night, and their wood-work was rent
and shattered; the house-door was violently shaken, and
great pieces of it were torn away; the gables of the
house were also pulled furiously to and fro.

One morning, before dawn, the old man we.t to the stable. An hour later his mistress rose, and, taking her milking-cans, followed him. As she reached the door of the stable, a terrible sound from within — the bellowing of the cattle, mingled with the deep bell-notes of an unearthly voice — sent her back shrieking to the house Thorhall leaped out of bed, caught up a weapon, and hastened to the cow-house. On opening the door, he found the cattle goring each other. Slung across the stone which separated the stalls was something. Thorhall stepped up to it, felt it, looked close: it was the cowherd, perfectly dead, — his feet on one side of the slab, his head on the other, and his spine snapped in twain.

The bonder now moved with his family to Tunga: it was too venturesome living during the midwinter night at the haunted farm ; and it was not till the sun had returned as a bridegroom out of his chamber, and had dispelled night with its phantoms, that he came back to the Vale of Shadows. In the mean time, his little girl's health had given way under the repeated alarms of the winter; she became paler every day : with the autumn flowers she faded, and was laid beneath the mould of the churchyard in time for the first snows to spread a virgin pall over her small grave.

At this time Grettir — of whom I have so often spoken — was in Iceland; and, as the hauntings of this vale were matters of gossip throughout the district, he heard of them, and resolved on visiting the scene. So Grettir busked himself for a cold ride, mounted his horse, and, in due course of time, drew rein at the door of Thorhall's farm, with the request that he might be accommodated there for the night.

"Ahem!" coughed the bonder: "perhaps you are not aware " —

"I am perfectly aware of all. I want to catch sight of the troll."

" But your horse is sure to be killed."

. "I will risk it. Glámr I *must* meet: so there's an end of it."

" I am delighted to see you," spoke the bonder: " at the same time, should mischief befall you, don't lay the blame at my door."

" Never fear, man."

So they shook hands. The horse was put into the strongest stable. Thorhall made Grettir as good cheer as he was able; and then, as the visitor was sleepy, all retired to rest.

The night passed quietly enough, and no sounds indicated the presence of a restless spirit. The horse, moreover, was found next morning in good condition, enjoying his hay.

" This is unexpected!" exclaimed the bonder gleefully. " Now, where's the saddle? we'll clap it on; and then good-by, and a merry journey to you."

" Good-by!" echoed Grettir. " I am going to stay here another night."

" You had better be advised," urged Thorhall. " If misfortune should overtake you, I know that all your kinsmen would visit it on my head."

" I have made up my mind to stop," said Grettir; and he looked so dogged, that Thorhall opposed him no more.

All was quiet next night: not a sound roused Grettir from his slumber. Next morning he went with the farmer to the stable. The strong wooden door was shivered and driven in. They stepped across it. Grettir called to his horse; but there was no responsive whinny.

"I am afraid"—began Thorhall. Grettir leaped in, and found the poor brute dead, and with its neck broken.

"Now," said Thorhall quickly, "I've got a capita' horse—a skewbald—down by Tunga. I shall not be many moments in fetching it. Your saddle is here, I think; and then you will just have time to reach"—

"I stay here another night," interrupted Grettir.

"I implore you to depart," said Thorhall.

"My horse is slain."

"But I shall provide you with another."

"Friend," answered Grettir, turning so sharply round that the farmer jumped back, half frightened, "no man ever did me an injury without ruing it. Now, your demon-herdsman has been the death of my horse. He must be taught a lesson."

"Would that he were!" groaned Thorhall; "but mortal must not face him. Go in peace, and receive compensation from me for what has happened."

"I *must* revenge my horse."

"An obstinate man must have his own way. But, .f you will run your head against a stone wall, don't be angry because you get a broken pate."

Night came on. Grettir ate a hearty supper, and was right jovial. Not so Thorhall, who had his misgivings. At bedtime the latter crept into his crib, which, in the manner of old Icelandic beds, opened out of the hall as berths do out of a cabin. Grettir, however, determined on remaining up: so he flung himself on a bench, with his feet against the posts of the high seat, and his back against Thorhall's crib; then he wrapped one lappet of his fur coat round his feet, the other about his head, keeping the neck-opening in front of his face, so that he could look through into the hall.

There was a fire burning on the hearth,—a smouldering heap of red embers. Every now and then a twig flared up and crackled, giving Grettir glimpses of the rafters, as he lay with his eyes wandering among the mysteries of the smoke-blackened roof. The wind whistled softly overhead. The clear-story windows, covered with the amnion of sheep, admitted now and then a sickly yellow glare from the full moon, which, however, shot a beam of pure silver through the smoke-hole in the roof. A dog without began to howl: the cat, which had long been sitting demurely watching the fire, stood up with raised back and bristling tail, then darted behind some chests in a corner. The hall-door was in a sad plight. It had been so riven by the vampire, that it was made firm by wattles only; and the moon glinted athwart the crevices. Soothingly the river prattled over its shingly bed as it swept round the knoll on which stood the farm. Grettir heard the breathing of the sleeping women in the adjoining chamber, and the sigh of the housewife as she turned in her bed.

Click, click! It is only the frozen turf on the roof cracking with the intense cold. The wind lulls completely. The night is very still without.

Hark! a heavy tread, beneath which the snow crackles. Every footfall goes straight to Grettir's heart. A crash on the turf overhead! By all the saints in paradise, the vampire is treading on the roof!

For one moment the chimney-gap is completely dark-
ened; the monster is looking down it; the flash of the
red ash *is* reflected in the two lustreless eyes. Then
the moon glances sweetly in once more, and the heavy
tramp of Glámr is audibly moving towards the farther
end of the hall.   A third — he has leaped down!   Grettir feels the board at his back quivering; for Thorhall
is awake, and is trembling in his bed.   The steps pass
round to the back of the house, and then the snapping
of wood shows that the creature is destroying some
of the outhouse-doors.   He tires of this, apparently; for
his footfall comes clear towards the main entrance to the
hall.   The moon is veiled behind a watery cloud; and,
by the uncertain glimmer, Grettir fancies that he sees
two dark hands thrust in above the door.   His appre-
hensions are verified; for, with a loud snap, a long strip
of panel breaks, and light is admitted.   Snap, snap!
another portion gives way, and the gap becomes larger.
Then the wattles flip out of their laces; and a dark arm
rips them out in bunches, and flings them away.   There
is a cross-beam to the door, holding a bolt, which slides
into a stone groove.   Against the gray light Grettir sees
a huge black figure heaving itself over the bar.   Crack !
that has given way, and the rest of the door falls in
shivers to the earth.

"O God!" exclaimed the bonder.

Stealthily the dead man creeps on, feeling at the beams
as he comes; then he stands in the hall, with the fire-
light on him.   A fearful sight!—the tall figure distended
with the corruption of the grave; the nose fallen off;
the wandering, vacant eyes, with the glaze of death on

them; the sallow flesh patched with green masses of decay: the wolf-gray hair and beard have grown in the tomb, and hang matted about the shoulders and breast; the nails too — they have grown. It is a sickening sight, — a thing to shudder at, not to see.

Motionless, with no nerve quivering now, Thorhall and Grettir held their breath.

Glámr's lifeless glance strayed round the chamber: it rested on the shaggy bundle by the high seat. Cautiously he stepped towards it. Grettir felt him groping about the lower lappet, and pulling at it. The cloak did not give way. Another jerk: Grettir kept his feet firmly pressed against the posts, so that the rug was not pulled off. The vampire seemed puzzled: he plucked at the upper flap, and tugged. Grettir held to the bench and bed-board, so that he was not moved himself: but the cloak was rent in twain; and the corpse staggered back, holding half in its hands, and gazing wonderingly at it. Before it had done examining the shred, Grettir started to his feet, bowed his body, flung his arms about the carcass, and, driving his head into the chest, strove to bend it backward, and snap the spine. A vain attempt! The cold hands came down on Grettir's arms with diabolical force, riving them from their hold. Grettir clasped them about the body again; then the arms closed round him, and began dragging him along. The brave man clung by his feet to benches and posts; but the strength of the vampire was greater. posts gave way: benches were heaved from their places; and the wrestlers at each moment neared the door. Sharply writhing loose, Grettir flung his hands round a roof-

beam. He was dragged from his feet; the numbing
arms clinched him about the waist, and tore at him,
every tendon in his breast was strained; the strain under
his shoulders became excruciating; the muscles stood out
in knots. Still he held on: his fingers were bloodless;
the pulses of his temples throbbed in jerks; the breath
came in a whistle through his rigid nostrils. All the
while, too, the long nails of the dead man cut into his
side, and Grettir could feel them piercing like knives
between his ribs. Ah! his hands gave way; and the
monster bore him reeling towards the porch, crashing
over the broken fragments of the door. Hard as the
battle had gone with him in-doors, Grettir knew that it
would go worse outside: so he gathered up all his
remaining strength for one final, desperate struggle.

I told you that the door had shut with a swivel into
a groove: this groove was in a stone which formed the
door-jamb on one side; and there was a similar block
on the other, into which the hinges had been driven.
As the wrestlers neared the opening, Grettir planted
both his feet against the stone-posts, holding Glámr by
the middle. He had the advantage now. The dead
man writhed in his arms, drove his talons into Grettir's
back, and tore up great ribbons of flesh; but the stone-
jambs held firm.

"Now," thought Grettir, "I can break his back;"
and thrusting his head under the chin, so that the
grizzly beard covered his eyes, he forced the face from
him, and the back was bent as a hazel-rod. "If I can
but hold on!" thought Grettir: and he tried to shout
for Thorhall; but his voice was muffled in the hair of
the corpse.

Crack! One or both of the door-posts gave way. Down crashed the gable-trees, ripping beams ard rafters from their beds; frozen clods of turf rattled from the roof, and thumped into the snow. Glámr fell on his back, and Grettir staggered down on top of him. The moon was at her full: large white clouds chased each other across the sky; and, as they swept before her disk, she looked through them like a pale saint in tribulation, with a brown halo round her. The snow-cap of Jorundarfell, however, glowed like a planet: then her white mountain-ridge was kindled; the light ran down the hillside; the bright disk started out of the veil, and flashed at this moment full on the vampire's face. Grettir's strength was failing him; his hands quivered in the snow; and he knew that he could not support himself from dropping flat on the dead man's face, eye to eye, lip to lip, nose to where the nose *had* been. The eyes of the corpse were fixed on him, lit with the cold glare of the moon. His head swam as his heart sent a hot stream through his brain. Then a voice from the gray lips said, —

" Thou hast acted madly in seeking to match thyself with me. Now learn that henceforth ill luck shall constantly attend thee; that thy strength shall never exceed what it now is; and that by night these eyes of mine shall stare at thee through the darkness till thy dying-day, so that for very horror thou shalt not endure to be alone."

Grettir at this moment noticed that his dirk had slipped from its sheath during the fall, and that it now lay conveniently near his hand. The giddiness which had oppressed him passed away : he clutched at the

13

sword-haft, and with a blow severed the vampire's throat; then, kneeling on the breast, he hacked till the head came off.

Thorhall came out now, his face blanched with terror: but, when he saw how the fray had terminated, he assisted Grettir gleefully to roll the corpse on top of a pile of fagots which had been collected for winter fuel. Fire was applied; and soon, far down Vatnsdalr, the flames of the pyre startled people, and made them wonder what new horror was being enacted in the Vale of Shadow.

Next day the charred bones were conveyed to the cairn and there buried.

CHAPTER XI.

RAED waked me at five the next morning. Kit was up loading our shot-gun. Wash was dressing.

"What's going on?" I inquired.

"Sluggard!" exclaimed Kit in tones of reproof, "hast forgotten?"

"Forgotten what?" for I was still half dead with sleep.

"Ingrate!" he hissed, pounding down the powder-wad, "hast forgotten the birthday of our nation, the glorious Fourth?"

It *was* the Fourth of July, sure enough.

"Hush!" Raed interrupted as I was beginning to disclaim any indifference on my part. "You are but a sleepy patriot, to say the best of you."

"Patriot!" cried Wash. "You don't call him a patriot, I hope! Patriot, indeed!"

"I think," Kit observed, capping the gun, "that we

195

ought to make this miserable traitor take the oath of allegiance on this our country's natal morning."

This aggravating talk had the intended effect, — to wake me up thoroughly. These uncompromising comrades of mine have long since discovered how to get my eyes open of an early morning. It is a daily wonder with me how we ever contrive to live so harmoniously together. There must be some strong bond of union between us, — strong and firm enough to span the chasm in our politics. Perhaps this is chiefly from the similarities of our tastes and present aims. But I doubt whether it be right for me thus to acquiesce and remain silent when sentiments so antagonistic to the doctrines in which I have been educated are confidently and defiantly uttered. I doubt. I must shortly take the matter into consideration. One thing becomes more and more apparent: the longer I associate with these young Northerners, the less I feel to differ from them. This fact of itself alarms me; and, if I were not sure that their own extreme political opinions were not similarly softened, I would not remain another month in the North.

"Don't wake the Dane yet!" cautioned Raed. "He has neither part nor lot in this matter. Let him snooze 'neath the wing of tyranny till the thunder of free American powder shall startle him."

The family were not yet up. We stole along the narrow passage, out into the yard; when Kit discharged the gun. It set all the dogs inside barking; the ravens croaked; the sheep bleated. Out rushed young Havisteen in very scanty garments. The old farmer and his son followed him.

"Why, what's up?" demanded our Danish friend. "Don't you know it's the Fourth of July, Jan?" Wash demanded. "It's Independence Day!—that glorious day which all Americans always celebrate in all places, the broad world over."

At this juncture Kit let off another gun. The old Icelander dodged back : he evidently considered us very dangerously insane. Jan had heard of our national powder-day, and, turning, hastily and laughingly explained the nature of our mania: but the old gentleman continued to regard our movements with astonishment; though a smile presently crept into his good-humored face, and he turned to re-explain the whole thing to his goodwife and Indridi, whose visages we could get glimpses of far back in the passage.

Ten guns were fired, and three hearty cheers given, in which young Havisteen joined; though his *hurrah* was quite out of chord with ours. None but a native-born American can give a genuine *hurrah!*

The celebration over, we went in to breakfast, and then set off for the Dettifoss Cataract, distant fifteen miles. Biarni went with us to show us the way. Jan had never been to the Dettifoss. A ride of half an hour took us to the Krafla Jokul, which we passed to the left. This volcano has been in eruption since Biarni's remembrance. The plain, for many miles around it, is strewn with black cinders. Over these the hoofs of our ponies clinked and rattled as we galloped onward. The country was bare of vegetation. Only here and there were to be seen thin patches of grass. Eight or nine miles from the *byre* we crossed another sulphur-range, from the ridge of which we caught a glimpse of the Jokulsa, the largest river of Iceland.

The Dettifoss is a cataract formed by the fall of this stream into a lava-*jau*, or rift. The Jokulsa rises in the Vatna Jokul, at the south-west coast of the island, and, after a course of a hundred and twenty-five miles nearly north-east, flows into the Arctic Ocean at the north-east corner of the island. The Dettifoss itself is a little north of east from Reykjalith. We took our compass; but, on going over the cinder-strewn plain to the south-east of Krofla, the needle became violently agitated (from the presence of iron in the clinkers, probably), and continued but a doubtful guide all the remainder of the day.

From the sulphur-hills we descended to a small lake, which Biarni called *Eylifr* (so Jan spelled it for us). A little way up the neighboring hillside we saw a shepherd's *byre* in one of the most lonely and desolate spots I ever saw inhabited. Why will folks cling to such utterly barren and profitless tracts, when the earth abounds with fertile dales and pleasant prairies? In reply to this query, Jan repeated the well-known Iceland proverb, "Hinn besta land sem solinn skinuar uppa," to the effect that "Iceland is the best land the sun shines upon."

Well, it's well they think so, I suppose; for the poor wretches could never get cash enough together to emigrate to the United States.

An hour later, while pacing over an utterly bare and fire-smitten tract, Biarni pointed off to the northward, where we saw a great misty column rising lazily toward the sky.

"Dettifoss," he said.

"Dettifoss," echoed Jan.

"*Hah-haw, hah-haw!*" to the sweating little horses; and away we go again, *cloperty clop*, leaping recklessly over and among the sharp black rocks. *Aiblains!* it makes me shudder to think how we used to ride over those lava-fields. Ere long, the heavy rumble of the falls began to be heard, and steadily increased during the next twenty minutes; till, on reaching a little patch of wild corn, Biarni gave the word to halt. Here we left our horses to refresh themselves, well persuaded that they were too tired to run away.

From this place a scramble of ten or fifteen minutes over rocks and ledges took us out to the river and fall. A feeling of disappointment came over me. It was not very high, nor yet very striking.

"No great shakes!" Wash said; and Raed at once pronounced it inferior to Godafoss. Young Havisteen looked puzzled, and not a little chagrined.

After letting us look at it for ten minutes or more, Biarni quietly made signs for us to follow him; which we did in some vexation. He kept on, winding amid crags and mighty bowlders which shut out the river. Indeed, I began to think he was taking us away from it; when on a sudden he turned to the right again, and, going in between two enormous rocks, stopped short on the brink of an abyss. We came up, and, lo! a second fall, infinitely grander, higher, and more stupendous! We glanced amazedly down, and then at Biarni. He looked perfectly honest and unmoved. Whether he had

really planned this surprise, or whether we had simply
and innocently sold ourselves, I know not. A river as
large as the Connecticut (for in Iceland one must not
judge of the volume of rivers by their length) falls into
a chasm in the dark basalt, full two hundred feet in
depth, at a single plunge! Such was the spectacle
before us. Why say more? The reader must imagine
the effect. So far as its volume goes, it much surpasses
Niagara in grandeur. The black chasm itself would be
a great natural curiosity. The thunderous plunge of
the cataract renders the scene strikingly complete. It
thrills and awes. So far as my observation goes, I place
the Dettifoss at the head of what I may conveniently
term the second-class cataracts.

We were seventeen hours making the round trip to
the falls. The farmer and family had retired ere we
got back. The tea-urn was still warm, and awaiting us,
however; and, after a light supper, we retired. It had
been one of the most fatiguing trips of our tour.

The next day was lowery and wet. We improved the
chance to have the farmer read the last chapters of the
Gretla. The reading and the interpretation occupied
the most of the day.

GRETTIR AT DRÁNGEY.

Poor Grettir! hustled from pillar to post, hunted
from one retreat to another, he had spent fifteen years
of hardship such as few men have undergone; yet the
hatred of his deadly foe, Thorir, had not expended itself.

DETTIFOSS.

At length, finding that no corner of Iceland was safe, he asked Gúthmundr the wealthy to advise him whither he should flee to be safe from his pursuers.

"There is only one spot, that I know of, where you can be in perfect security."

Grettir replied, that he had hitherto found no such spot.

Gúthmundr continued: "There is an islet in the Skagafjord, hight Drángey, abounding in fish and fowl; and no one can ascend it except by a rope-ladder which hangs down on one of its sides. If you can reach that spot, then you may be assured that it is in no man's power to touch you, so long as you are safe and sound, and able to guard the ladder."

"I will venture out there," said Grettir; "yet I am so timorous in the dark, that, to save my life, I cannot abide alone."

Gúthmundr answered, "Maybe; but I advise you to trust no one but your own self."

Grettir thanked him for his advice, and then hastened to his mother, at Bjarg, in the middle frith. The fear of the dark, to which he alluded, had come on him ever since his wrestle with Glámr, but had increased considerably of late. No sooner did darkness set in than the terrible eyes of the vampire seem to stare at him from the gloom. He slept lightly, starting in his dreams, and waking repeatedly during the night. This was undoubtedly brought on by the unceasing strain on his mind, and the excitability of nerves, caused by the hourly peril in which he had been living for so many years.

On his arrival at Bjarg his mother greeted him affectionately, and told him that she would indeed be glad if he could remain with her; though she feared it would be too venturesome to do so, as Thorir would certainly discover his retreat before many days had elapsed.

The outlaw replied, that he would give her no inconvenience: "For," said he, "I care to take no more trouble about preserving my life. I can bear my solitude no longer." He then told his mother of Gúthmundr's advice; adding, that he would try his best to reach Drángey, but that he must endeavor to secure some trustworthy companion to be with him.

Illugi, his brother, now fifteen years old, a fine, noble boy, was present during the conversation; and at these words of Grettir he started up, caught his hand, and said, —

"Brother, I will go with you if I may; though I fear you will look on me as but a feeble helpmate: yet I will be faithful to you, and stand by you to the last."

Grettir answered, "Of all men, my brother, I would rather have you with me; and willingly will I consent to your joining your lot with mine, if our mother has no objection."

"Sorrows never come singly," replied the aged woman. "I can hardly bear to part with Illugi: yet I know how dire is your necessity of a comrade, son Grettir; therefore I will not be selfish and keep him. It costs me a bitter pang to part with both my sons in one day."

Illugi was delighted at having thus easily obtained

that on which he had set his heart; and he thanked his mother cordially.

The mother provided her sons with money, and such chattels as they would require on the island; and then she accompanied them outside the farm-yard, and, before parting with them, said, "Farewell, my two brave boys! I know that I shall never see you again; but what will befall you in Drángey I know not. Only of this I am certain, — that there you will die; for many will resent your occupation of that island: my dreams have long forewarned me that you will not be divided in your deaths. Beware of treachery: shun any dealings with sorcery; for nothing is more powerful than witchcraft. My blessing be upon you both!" She could speak no more, for her voice was choked with sobs: so, sitting down on a stone, she covered her eyes with her hands; and the tears trickled between her fingers, falling in bright drops on her lap.

"Do not weep, mother!" said Grettir. "What though we both die? It shall ever be said of you, that you bore sons, and not daughters. Long life and health attend you!"

Then they parted; and the brothers went north, and visited their kinsmen. So passed autumn; and with the approach of cold they went towards Skagafjord, crossed the Vatnskarth and Reykjaskarth to Langholl, and reached Glaumbœr at the close of day. Grettir had flung his hood over his shoulders, though the wind was piercingly cold; for it was not his wont, fair or foul, warm or cold, to wear any thing on his head.

Near the little farm just mentioned the brothers

stumbled upon a tall, thin man, dressed in rags, and with a very big head. They asked each other's names; and the fellow called himself Glaum: he was a bachelor out of work, and withal a gad-about, fond of strolling through the country picking up and retailing news. He was a terrible boaster; but most people thought him both a coward and a fool. He amused the brothers by his continual chatter and by the fund of gossip which he possessed. Grettir was especially pleased with him: and, when Glaum offered to be his servant, Grettir accepted him gladly; and the man became thenceforth his constant attendant.

Says Glaum, " It is a wonder to all the people hereabouts that you wear nothing on your head in such weather as this; and, i' faith ! it is no marvel that you are the man they take you for, if you do not mind the cold. Why, there were two of the bonder's sons down yonder going after the sheep, and they could not get clothes enough to put on them, so benumbed were they; and yet they are plucky fellows too ! "

After this they went to Reynines; thence they proceeded to the strand, where there is a little *byre*, Reykir, with a hot spring in the *tún*, belonging to a man named Thorwaldr. Grettir offered him a bag of silver if he would flit him across to Drángey by moonlight; and to this the man agreed.

On arriving at his destination, Grettir was well pleased with the spot; for it was covered with a profusion of grass, and was so precipitous, that it seemed impossible for any one to ascend it without the aid of the rope-ladder which hung from strong staples at the sum mit. In summer the place would swarm with sea-birds; and at that time there were eighty sheep left on the island for fattening.

One of the principal chiefs in the Skagafjord was Thorbjorn, nicknamed "The Hook,"—a hard-hearted, ill-disposed fellow. His father had married a second time, and there was no love lost between the step-mother and Thorbjorn. It is said that one day, as the Hook was sitting at draughts, she passed, and, looking over his shoulder, noticed that he had made a foolish move : so she laughed; whereupon Thorbjorn retorted angrily. She instantly snatched up a draught-man, and, laying it against his cheek-bone, pressed it into his eye, so that the ball started out of its socket. He sprang up with a curse, and dealt her such a blow, that she took to her bed, and died of the injury. Thorbjorn went from bad to worse; and, leaving home, he settled at Vithvik.

As many as twenty farmers had rights of pasturage on Drángey; but the Hook and his brother had the greatest share.

About the time of the winter solstice, the bonders busked them to visit the island and bring home their sheep. They rowed out in a large boat, and, on nearing the island, were surprised to see figures moving on the top of the cliffs. How any one had reached the islet without their knowledge was a puzzle to them; and they had not the slightest suspicion who these occupants could be. They pulled hard for the landing-place where hung the ladder; but Grettir drew it up before the boat stranded.

The bonders shouted to know who those were on the

crags; and Grettir, looking over, told his name and those
of his companions.

The bonders asked who had flitted them across to the
island. Grettir answered, "If you wish particularly to
know, I will tell you: it was a man with a good boat
and strong arms, and one who was rather my friend than
yours."

"Let us get our sheep," cried the bonders; "and, you
come to land with us, we will charge you nothing for
those of our sheep you have eaten, and we will let you
go from us in peace."

"Well offered," answered Grettir; "but he who takes
keeps hold, and 'a bird in the hand is worth two in the
bush.' Believe me, I never leave the island till I am
carried from it dead."

The bonders were silenced. It seemed to them that
they had got an ugly customer on Drángey, to get rid
of whom would be no easy matter: so they rowed home,
very ill pleased at the result of their expedition.

The news spread like wildfire, and was talked about
all through the neighborhood; but no one could devise
a plan for getting rid of the outlaw.

Winter passed; and, at the beginning of spring,
the whole district met at the "Thing," or Council of
Hegraness, — an extensive island at the mouth of the
Heradsvatn River. The gathering was thronged; and
the litigations and merry-making made the Thing
last over many days. Grettir guessed what was going
on by seeing a number of boats pass the head of the
fiord. He became very restless, and at last announced
to his brother that he intended being present at the

council. Illugi thought this sheer madness; but Grettir was resolute. He begged Illugi and Glaum to watch the ladder, and await his return.

Then he crossed to the mainland, and hastened in disguise to the council, where he found that sports of all kinds were going on among the able-bodied young men. Grettir was dressed in an old-fashioned suit, very dirty, and falling to tatters. He had on a fur cap, which was drawn closely over his eyes, and concealed his face, so that no one recognized him. He sauntered among the booths till he reached the spot where the games were taking place.

Among the wrestlers no man surpassed Thorbjorn Hook in skill and prowess. He threw all the strongest men of the neighborhood; and when he had cleared the ground of antagonists, and found that there was no one to oppose him, he stood still, and cast his eyes round him. Suddenly they rested on a tall fellow in the shabbiest and quaintest of suits, but who looked so strongly built, that Thorbjorn walked up to him, and caught him by the shoulders. But the man sat still, and he could not move him from his seat. "Well," exclaimed the Hook, "you are the first fellow I have seen for many a day whom I cou'dn't pull off his stool. Come now, and wrestle with me. Yet tell me, first, what is your name."

"Guest," answered the stranger.

"A welcome guest too," quoth the bully, "if you will wrestle with me."

The man replied that they would not be fairly matched, as he was little skilled in athletic sports.

Several men now chimed in, begging the stranger to try what he could do with Thorbjorn, or, at all events, with one of the others.

"Long, long ago," quoth he, " I was able to throw my man as well as the best of you; but those days are gone by, and now I am out of practice."

As he only half refused, the bystanders urged him all the more.

"Now mark you," said he: "I yield on one condition; and that is, that you take your oath to let me go free to my home without one of you lifting your hand against me."

There was a general shout of acquiescence; and Hafr, one of the number, recited the peace-oath in the following legal form : —

" Here set I peace among all men towards the man Guest, who sits before us; and in this peace I include all the priesthood-holders, and well-to-do bonders, and all the young weapon-bearing men, and all the men of the Hegraness district, whether present or absent, named or unnamed. These are to leave in peace, and give passage, without let or hinderance, to the aforenamed stranger, that he may sport, wrestle, make merry, abide with us, and depart from us, without stay, whether he may need to go by land or flood. He shall have peace in all places, named or unnamed, as long as is necessary for him to reach home with ease: so long only shall peace last.

"I set this reconciliation between him and us, our relations, our friends, and kinsmen, male or female, free or thrall, child or full-grown. May the breaker of this peace, and breaker of this oath, be cast out of the presence of God and good men, from the heavenly kingdom, from the company of the saints and just men! Let him be an outcast from land to its farthest limits, — far as men chase wolves at farthest, as Christians seek churches, as heathens sacrifice in shrines, as flame burns, earth produces, as baby calls its mother, and mother bears baby, as fire is kindled, ships glide, lightnings flicker, sun shines, snow lies, Finns slide on snow-shoon, fir grows, falcon flies in the spring day with a fair breeze under its wings, far as heaven bends, earth is peopled, winds sweep waters to the sea, churls grow corn! He shall be banished from churches and the company of Christian men, from heathen folk, from house and den, from every home — save hell! Now let us be at-oned and agreed, each with each, in good will, whether we meet on mountain or shore, on ship or on skate, on ground or glacier, at sea or in saddle; as friend meets friend on the deep, as brother meets brother abroad, let us be at-oned one with another; as father with son, as son with father, in all our dealing. Lay we now hand to hand, and hold we now true peace, and keep we every word spoken in this our peace-telling, before God and good men, and all those who hear my words and stand around."

After a little hesitation, the oath was taken by all

Then said Guest, "Now you have done well: only beware of breaking your oaths. I am ready on my part, without delay, to fulfil your wishes." Then he flung aside his hood and almost all his tatters.

The assembled chiefs looked at each other, and were rather disconcerted; for they saw that there stood before them the redoubted Grettir, Asmund's son. They were

14

silent; and Hafr thought that he had acted somewhat rashly. The throng broke up into knots, and began to discuss whether the oath should be kept or not.

"Come, now," shouted Grettir, "let me know your purpose; for I shall not long sit naked. There is more danger to you than to me in the breach of your oaths."

He got no answer; but the chiefs moved away to discuss the question. Some wanted to break the truce; others wanted to keep it. Then Grettir sang, —

> " Many trees of wealth,* this morning,
> Failed the well-known, well to know.
> Two ways turn the sea-flame branches †
> When a trick on them is tried.
> Falter folk their oath fulfilling;
> Hafr's talking lips are dumb."

Said a man hight Tongue-stone, "You think so, do you, Grettir? Well, you are a man of dauntless courage: I will say that for you. Look, now! the chiefs are in deep consultation about what is to be done with you."

Then Grettir sang, —

> " Lifters of shields ‡ rub their noses,
> Shield-tempest gods ‡ shake their beards,
> Fierce-hearted serpents'-lair-scatterers §
> Go on their way, much regretting
> Peace they have made, now they know me."

---

* Periphrasis for men.
† Sea-flame = gold; and sea-flame branches = warriors.
‡ Periphrasis for warriors.
§ Serpents' lair = gold; serpents'-lair-scatterers = men.

Then out spake Hjalti of Hóf, brother of Thorojorn Hook, "Never let it be said of us that we break an oath, even though it were inconsiderately taken. Grettir shall be at full liberty to go to his home in peace; and woe betide him who lays hand on him to do him an injury! But, should he venture again ashore, we are free from our oath."

All except Thorbjorn Hook agreed to this, and were glad that Hjalti had spoken out as became a chieftain.

The wrestling began by Grettir being matched with Thorbjorn; and, after a short struggle, Grettir freed himself from his antagonist, leaped over his back, caught him by the belt of his trousers, lifted him off his legs, and flung him over his back.

It was next proposed that Grettir should be matched against the two brothers together; and he readily agreed to this. The wrestling continued with unabated vigor, and it was impossible to tell which side had the mastery; for, though Grettir repeatedly threw one brother after the other, yet he was unable to hold them both down at the same time. After that all three were covered with blood and bruises, the match was closed by the judges deciding that the two brothers conjointly were not stronger than Grettir alone; though they were each of them as powerful as two ordinary abled-bodied men.

Grettir at once left the Thing, rejecting all the entreaties of the farmers that he should leave Drángey; and, on his return to the little island, he was received by his brother Illugi with open arms.

The small bonders began to feel seriously their want of the island for autumn-pasture; and, as there seemed

no prospect of their getting rid of Grettir, they sold
their rights to Thorbjorn Hook, who set himself in
earnest to devise a plan by which he could possess himself of the island.

When Grettir had been two winters on the island, he
had eaten all the sheep except one ram, — a piebald
fellow, with magnificent horns, which became so tame,
that every evening he came to the hovel which Grettir
had erected, and butted at the door till he was admitted.
The brothers liked their place of exile, as there was no
dearth of eggs and birds; besides which, a considerable
amount of drift-timber was thrown upon the strand, and
served as fuel.

Grettir and Illugi spent their days in clambering
among the rocks, and rifling the nests; and the occupation of the thrall was to collect drift-wood, and keep up
the fire in the cottage.

The churl lost his spirits, and became idle, morose, and
reserved. One night, notwithstanding Grettir's warn-
ings to him to be careful, as they had no boat, he let the
fire go out. Grettir was very angry, and told Glaum
that he deserved a sound thrashing for his neglect. The
thrall replied that he was heartily tired of the life he
had been leading on the island, being scolded or beaten
whenever any thing went amiss.

Grettir asked Illugi what had better be done; and his
brother replied, that the only thing for them to do was to
await the arrival of a boat from the friendly farmer at
Reykir.

"We shall have to wait long enough for that," said
Grettir: "our only chance is for me to swim ashore and
procure a light."

"For Heaven's sake!" exclaimed Illugi, "do not attempt any thing of the kind; for we are undone if any thing happens to you."

"Never fear for me," said Grettir: "I was not born to be drowned."

From Dríngey to Reykir is about four miles. Grettir prepared for swimming by dressing in loose, thin drawers, and a sealskin hood: he also tied his fingers together, that they might offer more resistance to the water when he struck out.

The day was warm and fine. Grettir started in the evening, when the tide was in his favor; whilst his brother anxiously watched him from the rocks. At sunset the outlaw reached Reykjaness, after having floated or swum the whole distance. Immediately on coming to land he went to the warm spring, and bathed in it before entering the house. The door of the hall was open, and Grettir stepped in. A large fire had been burning on the hearth, so that the room was very warm. Grettir was so thoroughly exhausted with his swim, that he lay down beside the hot embers, and was soon fast asleep. In the morning he was found by the farmer's daughter, who gave him a bowl of milk, and brought her father to him. Thorwald furnished him with fire, and rowed him back to the island, astonished beyond measure at his achievement in having swum such a distance.

The inhabitants of Skagafjord were angry with Thorbjorn Hook for not having rid the island of its tenants, notwithstanding all his fine promises; but Thorbjorn was sorely puzzled to know what measures to take.

During the summer a ship arrived in the frith, commanded by a young, active fellow, Hœring by name, who was famous for his skill in climbing. He lodged with Thorbjorn during the autumn, and was continually urging his host to row him out to Drångey, that he might escalade the precipitous sides of the islet. Thorbjorn required very little pressing; and one fine afternoon he flitted his guest out to the island, and put him stealthily ashore, without attracting the notice of those on the height.

On reaching the usual landing-place, which was on the opposite side of the island, Thorbjorn shouted, and brought Grettir and his brother to the verge of the cliff. The old arguments were repeated to persuade Grettir to come to the mainland, and with the usual success. The Hook, however, succeeded completely in his attempt to withdraw the outlaw's attention from the farther side of the islet, up which Hœring was clambering.

The young merchant reached the top by a way never attempted before nor since ; then, pausing only to take breath, he advanced towards the brothers, who were leaning over the verge of the cliff, little dreaming of danger in their rear.

Grettir was engaged in angry altercation with the Hook : but the young brother took no part in the conversation ; and, beginning to feel weary of his position, he turned on one side to relieve his elbows, which had rested on the rock. In so doing, he caught sight of Hœring.

"Brother, brother !" exclaimed he, "here comes a man towards us, brandishing an axe, and bent on mischief !"

"Go after him yourself, lad !" said Grettir : "I will guard the ladder !"

Illugi sprang up, and rushed towards the young merchant, who at once took to flight, ran to the edge of the
crag, leaped over, and was dashed to pieces among the
rocks. That spot is called "Hœring's Leap" to this
day.

"Now, Thorbjorn," shouted Grettir when Illugi returned and told him what had taken place, "you had
better row round to the other side of the isle, and gather
up the remains of your friend."

The Hook pushed off from the strand, and returned
home, ill enough pleased with what had taken place;
and Grettir remained at Drángey unmolested through
the winter.

At this time died Skapti the lawgiver: and, in the
following spring, Grettir's relations and friends moved
for a repeal of his sentence of outlawry; but his enemies
opposed this vehemently, declaring that he had com-
mitted many crimes since he had been pronounced an
outlaw.

A new lawgiver, named Steinn, was elected; and,
when the case came before him, he gave his opinion,
that, after twenty years, the sentence became null and
void: so that Grettir's kinsmen had every reason for
hoping that in two years, when he would have completed
the twenty, he would be restored to their society.

Thorbjorn Hook was exasperated beyond measure at
the prospect of Grettir slipping through his fingers
after all; and he returned from the Thing, brooding over
fresh schemes against the outlaw.

It happened that he had an old feeble foster-mother,
—a woman of malicious disposition; and, when Thorb-

jorn could get help nowhere else, he came to her, as in her youth she had dabbled in sorcery, but had long ceased to practise it, when, after the introduction of Christianity, it became illegal, and was punishable with banishment. However, as the old saw has it, "What is learned in youth is remembered in age;" and though the old woman was believed to have forgotten her witchcraft, yet it remained stored up in the chambers of her mind.

"Ah!" said she when Thorbjorn came to her, "I see that as a last resource you come to me, a bed-ridden old woman, and ask my help. Well, I will assist you to the best of my power on one condition; and that is, that you yield me implicit obedience."

The Hook answered her that he was quite willing to consent, as he had long since learned to rely on his foster-mother's advice, as being most salutary.

When the month of August came round, the hag said to her foster-son one beautiful day, "The sea is calm, and the sky bright: what say you to our rowing over to Drángey, and stirring up the old quarrel with Grettir? I will accompany you, and listen to what he says: I shall then be able to judge what lot awaits him. Besides, I can death-doom him as I please."

The Hook answered, "I am tired of going to Drángey; for I never return from it a whit the better off than when I started."

"Remember your promise," said the old woman. "I shall have nothing to do with you unless you follow my advice."

"Well then, foster-mother," quoth Thorbjorn, "let us

go; though I vowed that my third visit should be the death of Grettir."

"Have patience," said the hag: "time and trouble are needed before that man is laid low. And what the result will be I know not: it may be your gain, and it may be your ruin."

Thorbjorn ran out a long boat, and entered it with twelve men: the hag sat in the bows, coiled up amongst wraps and rugs.

When they reached the island, the brothers ran to the ladder; and Thorbjorn asked whether Grettir was yet tired of his island.

Grettir replied as he had replied before, "Do what you will: in this spot I await my destiny."

Thorbjorn saw now that his journey was likely to be without avail. "I see," said he, "that I have to do with the worst of men. One thing is clear enough: it will be a long time before I pay you another visit."

"So much the better," answered Grettir: "I shall not count it as a misfortune if I never see you again."

At this moment the hag began to stir in the bows of the boat. Grettir had not previously observed her presence. Now with a shrill voice she cried, "These men are sturdy; but luck has deserted them. See what a difference there is between folk! You, Thorbjorn, make them good offers, which they foolishly reject: those who refuse good when it is offered them always come to a bad end. Grettir, I wish you to be lost to health, wisdom, luck, and prudence! May these blessings be constantly on the wane the longer you live! and may your days henceforth be fewer and sadder than those preceding them!"

As she spoke a cold shudder ran over Grettir's limbs
and he asked what fiend that was in the ship. Illugi
replied that she must be the foster-mother of Thorbjorn.

"Since an evil fiend is with our foes, we can expect
nothing but the worst," said Grettir. "Never before
have I been so agitated at words spoken as whilst the
hag was pouring forth her curses on me. I know now
that evil must befall me from her witchcraft; but she
shall have a reminder of her visit to me." Then he
snatched up a large stone and flung it into the boat, so
that it fell upon the bundle of rugs, among which lay
the aged woman. As it struck, there rose a wild shriek
from the witch; for the stone had fallen on her leg, and
snapped it asunder.

"Brother, you should not have done this," said
Illugi.

"Blame me not," answered Grettir. "I only wish
that the stone had fallen on her skull, and that her life
had been sacrificed instead of ours."

On the return of Thorbjorn to the mainland, the hag
was put to bed, and the Hook was less pleased than
ever with his trip to the island.

"Be not downcast," said his foster-mother: "this is
the turning-point of Grettir's fortunes, and his luck will
leave him more and more. I have no fear of not having
my revenge, should my life be spared."

"You are a resolute woman, foster-mother," said
Thorbjorn Hook.

After a month the old woman was able to leave her
bed and limp across the room. She one day demanded

to be led down to the shore. Her wishes were complied
with ; and, on reaching the strand, she hobbled up and
down till she found a large piece of drift-timber, just
large enough for a man to carry upon his shoulder. Then
she ordered it to be rolled towards her, and turned over.
She examined it attentively. The log seemed to have
been charred on one side ; and this burned portion she
ordered to be planed away : then, taking a knife, she cut
runes on it, and smeared them with her blood, chant-
ing over them, as she limped round the beam, a wild
spell, that it might be borne to Drángey, and there work
Grettir's ill. The piece of timber was then pushed into
the waves, and thrust off from shore. A fresh northerly
wind was blowing ; but the beam swam against wind and
tide, and held on its course direct for the outlaw's isle.

The old witch returned to Vithvik. Thorbjorn did
not think that any thing would come of what she had
done ; but she bade him be of good cheer, and wait till
she gave him fresh orders.

In the mean time, Grettir, his brother, and the churl
were on Drángey, catching fish and fowl for their winter-
supplies.

The day after that on which the hag had charmed the
piece of timber, the two brothers were walking on the
strand to the west of the island, looking for drift-wood.

" Here is a fine log ! " exclaimed Illugi : " help me to
lift it on my shoulder, and I will carry it home."

Grettir spurned the beam with his foot, saying, "I do
not like the looks of it, little brother. Runes are cut on
it, and they may betide us ill : who knows but this log
may have been sent hither for our destruction ?"

Then they sent it adrift; and Grettir warned his brother not to bring it to their fire.

They returned in the evening to their hovel, and did not mention the matter before the thrall.

The next day they found the same beam washed up, not far from the foot of the ladder. Grettir thrust it out to sea again, saying that he hoped he had seen the last of it.

The weather began to break up; and several days of storm and rain succeeded each other, so that the three men remained in-doors till their stock of fire-wood was nearly expended.

Then they ordered Glaum to search the shore for fuel. The fellow started up with an angry murmur, and left the room, muttering that the weather was too bad for a dog to be sent out in it. Then he went to the rope-ladder, descended it, and found the same beam cast up at its very foot.

Rejoiced at having so soon obtained what he wanted, he threw it over his shoulder, strode with it to the hut, and flung it down by the door.

Grettir heard the sound; and, springing up, he exclaimed, "Glaum has got something at last! Let us see what he has found!"

Then, taking his axe, he went outside.

"Now," says Glaum, "you chop it up, as I have had all the trouble of bringing it."

Grettir was angry with the fellow; and, without paying much attention to the log itself, he brought his axe down on it with a sweep. The blade struck, glided off, and cut into Grettir's right leg, below the knee, with such force that it stuck in the bone.

Grettir looked at the beam, and, recognizing it at once, said, "The worst is at hand. Misfortunes never come singly. This is the very log which I have rejected twice. Glaum, you have done us two ill turns, — first in letting out the fire, secondly in bringing home this accursed beam; and, if you commit a third, it will be the death of you."

Illugi bound up his brother's wound with a rag. There was but little flow of blood; but it was an ugly gash.

Grettir slept well that night. For three days and nights he was without pain; and the wound seemed to be healing nicely, and skin to be forming healthily over it.

"Well, brother," said Illugi, "I think that this cut will not trouble you long."

"I hope not," answered Grettir; "yet I have my fears."

On the fourth evening they laid them down to sleep as usual. Towards midnight the lad Illugi awoke, hearing Grettir tossing about in his bed as though in pain.

"Why are you so restless?" he asked.

Grettir replied that he felt great anguish in his leg, and that he thought some slight change must have taken place in the wound.

The boy blew the embers on the hearth into a flame, and by its light examined his brother's leg. He found that the foot was swollen and purple, and that the wound had re-opened, and looked far more angry than when first made.

Intense pain followed, so that the poor outlaw could not remain quiet for one moment; and sleep no more visited his eyes.

Illugi remained by him, continually holding his brother's hand, or bringing him water to slake his unquenchable thirst.

"We must prepare for the worst," said Grettir. "This sickness is the result of sorcery. The hag is revenging on me that stone which I cast at her."

Illugi replied, "I ever thought evil would come of it."

"What is done cannot be undone," said Grettir; and then sitting up, supporting himself against his brother's breast, he sang, —

> I fought with sword in bright old days, —
>   In days when I was young, —
> When gladsome song and roundelay
>   From happy heart I flung.

> I fought with sword in bright old days,
>   When earth to me was fair,
> And, fresh as heart, the lightsome breeze
>   Did toss my yellow hair.

> I fought with sword in bright old days,
>   I loved the merry clang,
> When brand met brand, and shield met shield,
>   And axe on helmet rang.

> As now I chant of youthful days
>   In fitful broken rhyme,
> I seem to hear from my blue blade
>   A wild war-music chime.

> I lowly laid the robber-band;
>   I rescued wife and maid :
> My haft and hilt were purple-dipped,
>   And purple war my blade.

And, when my friends for fire did pray,
    I sought it past the wave;
Though 'neath me gaped the fiord dark, —
    Dark as an open grave.

When I returned to seek old home,
    I found my kinsmen dead;
I was a banished, outlawed man;
    A price was on my head.

A hunted man by night and day,
    On mountain, moor, and fen;
For eighteen years to shun and flee
    The face of fellow-men! —

For eighteen bitter years to bear
    Fasting and cold and pain,
And never know, when I lay down
    If I should wake again!

And now, coiled up with fevered blood,
    A grim old wolf I die;
Whilst dripping skies above me spread
    And winds sob sadly by.

O'er tired heart and drowsy head
    Does welcome slumber creep,
As little babe on mother's knee
    Will softly drop asleep.

With folded feet and closèd palms,
    I will not stir nor wake,
But, hushed in happy dreaming, lie
    Till the last morning break.

And, if men ask who lieth thus,
    Say, " 'Tis a tired breast,
Now finding peace, finding calm,
    Finding rest."

"Let us be cautious now," said Grettir; "for Thorbjorn will make another venture. Glaum, do you watch
the steps by day, and draw them up at dusk. Be a
faithful servant to us; for much depends on your fulfil-
ling your duty: and I forewarn you, that, if you betray
your trust, it will cost you your life."

Glaum promised well.

The weather daily became worse; and a fierce, north-
east wind blustered over the country, bearing with it
cold and sleet, and powdering the highlands with snow.
Grettir asked nightly whether the ladder had been
drawn up. Glaum answered churlishly, "How can you
expect people to come out in such a storm as this? Do
you think that folk are so anxious to kill you, that they
will be crazy enough to jeopardize their own lives in the
attempt? No, no! You have lost all your pluck and
manliness since you have been a little unwell. You
are now scared and frightened at the merest trifles."

Grettir answered, "You have none of our pluck and
manliness yourself! Go, now, and guard the ladder, as
you have been bidden, instead of standing here reproach-
ing us with cowardice!"

So Illugi and his brother drove the churl from the
house every morning, notwithstanding all his angry re-
monstrances.

The pain became more acute, and the whole leg
became inflamed and swollen. Signs of mortification
appeared, and wounds opened in different parts of the
limb, so that Grettir felt that the shadow of death was
upon him. Illugi sat night and day with his brother's
head on his shoulder, bathing his forehead, and doing

his utmost to console the fleeting spirit. A week had elapsed since the wound had been made.

Thorbjorn Hook was at home, ill pleased at the failure of all his schemes for dispossessing Grettir of the island. One day his foster-mother came to him, and asked whether he were ready now to pay the outlaw his final visit. Thorbjorn replied that he had no wish to do so, as it would come to nothing; and asked his foster-mother whether she had any desire to seek out Grettir again, or whether she had been satisfied with the success of her former visit.

"I may not seek him myself," answered the hag: "but I have sent him my greeting, and by this time it has reached him. Speed now to Drángey as swiftly as you can row; for, if you delay, he will be beyond your reach."

The Hook had come off so ignominiously on every former occasion when he had visited the island, that he did not much relish the notion of making another attempt, especially on a day when it would be dangerous to venture on the water in a boat.

"You're a helpless fellow!" exclaimed his foster-mother when Thorbjorn told her his objections to her scheme. "Do you think that I, who have called up this storm, cannot refrain it from doing you injury?"

Well, in the end, the man allowed himself to be persuaded: so he beat up the neighboring farmers, asking them to assist him in manning a large boat. None of them would come with him: but the Hook brought twelve of his own men; his brother Hjalti lent him three more; Eirik of the Good-dale sent him one man:
15

Tongue-stone furnished him with two; Halldorr let him have six of his house-churls; and these were all he could get. Of these, the only two whose names I need mention were Karr and Vikarr. Thorbjorn went with his party to Haganess, where he borrowed a large sailing-boat. None of the men were in good spirits, as the weather was so bad, and they had no confidence in their leader. By dusk they got the vessel afloat, and spread sail; and, with a lurch, she ran out to sea.

As the wind was from the north-east, they were under the lee of the high cliffs, and were not exposed to the violence of the gale.

Heavy scuds of rain and sleet swept the fiord: the sky was overcast with dense whirling masses of vapor; and, beneath their shadow, the waters of the frith were black as ink. For one moment the clouds were parted by the storm, and the rowers looked up to see the heavens barred with the crimson rays of the northern light; then the vapors, dense as volcanic smoke, swept across the gap. A flame ran along the cordage, and finally settled on the masthead of the boat, swaying and rocking with the motion of the vessel. It was that electric spark which Mediterranean sailors call "St. Elmo's Light;" and Icelanders, "*Hrævarelldr.*"

A line of white foam marked the base of Drángey; and now and then a great wave from the mouth of the fiord thundered against the crags, and shot in spouts of foam high into the air.

Along the western shore of the frith, which was exposed to the full brunt of the gale, the mighty billows were beaten into white, yeasty heaps of water, rolling on.wards and recoiling, shivering against 'he rocks, and falling back in lashing spray, booming down long caveins till they choked them, and then bursting out with a roar in steam-like jets. Upon the top of Drángey, one ruddy spark shone from the window of the hovel in which lay the dying outlaw; and it was reflected as a streak of fire on the tossing deep.

Gulls cried and wheeled around the smack as it ploughed its way through the water; and the stormy petrel fleeted past in the trough of the waves. The kittiwakes and tern wavered and dipped before the boat, uttering their melancholy scream, "Kreeah, kreeah!" The diver passed, dancing like a cork, or rushing through a wave to appear on the farther side with a fish in its beak. Seals rose out of the water, and watched the boat, floating with only their round black heads above the surface, — heads in the gloom appearing so fearfully like those of human beings, that it seemed to the shuddering rowers that the drowned men of the fiord had risen to greet them on their desperate errand.

Now let us return to Grettir.

He had been in less pain that day. Illugi had not left him, but remained faithful to his post.

The churl had been sent out as usual to watch the ladder, and draw it up at nightfall; but, instead of doing as he was bid, the fellow laid himself down at the head of the steps in a sheltered nook, and went to sleep. As dusk set in, the thrall partially awoke, and looked drowsily at the ladder. "Humph!" said he: "I see no use in taking the trouble of pulling this up to-night, when there is such a sea running that no

boat could venture out on it. I'll just take another
snooze, then saunter home and say that all is
safe." So he turned on his side, and was soon snoring.

When Thorbjorn and his party reached the shore,
they found that the ladder still hung down.

"We are in luck's way!" exclaimed the Hook.
"Now, my men, perhaps you will think that our journey
will not prove as bootless as you expected. Up the
ladder with you; and let us all be of good courage!"

Then they ascended, one after the other; Thorbjorn
taking the lead. On reaching the top, they noticed
Glaum, asleep under a rock, snoring loudly. Thorbjorn
recognized the man at once, and struck him over the
shoulders with his sword-hilt, bidding him wake up,
fool that he was, and tell them truly all that he knew
about those whom he sought.

Glaum turned over on his side, rubbed his eyes, and
growled forth, "Cannot you leave a poor wretch alone?
Assuredly never was man so ill treated before: you
won't even let me sleep out here in the cold!"

"Idiot!" exclaimed the Hook, "look up and see
who are come! We are your foes, who purpose slaying
every man of you!"

Glaum started up, and screamed with terror when he
saw the black figures around him.

"Silence!" cried Thorbjorn. "I give you your choice
of two things, — answer the questions I put to you, or
die on the spot."

The churl was silenced, and stood trembling before

the Hook, with great drops of perspiration rolling off his face.

"Are the brothers in the house?" asked Thorbjorn, "or shall we find them out of doors?"

"Oh!" cried Glaum, "they are both within, — Grettir sick to death, and Illugi watching and never leaving him!"

The Hook asked for particulars; and Glaum told him all the circumstances of Grettir's being wounded. Then the Hook burst out laughing, and said, "The proverbs come true, — 'Old friendships are the last to be broken;' and 'Woe to him who has a thrall for an only friend!' especially if he be such a fellow as you, Glaum; for shamefully have you betrayed your master, bad though he be!"

Some of the men caught Glaum by the throat, and beat him till he was nearly senseless; then they flung him down, and pushed on towards the hovel.

In the mean time, Illugi had been sitting near the fire, with his brother's head on his lap; whilst Grettir lay on some sheepskins beside the hearth. All that evening the sick man's eyes had been wandering among the rafters, watching the light play among them as the firewood blazed up or smouldered away. Presently he turned his head towards his brother, saying that he thought he could sleep; and in a few moments he closed his eyes.

Illugi watched his face, kindled by the scarlet glow from the embers. It was more tranquil than he had seen it for many days: the muscles were relaxed; and the wrinkles furrowed on the brow by the intense pain which the poor outlaw had suffered were now smoothed quite away. Grettir's face was not handsome; but it was grave and earnest, tanned dark by continual exposure to the weather. His breath came evenly in sleep. One hand lay open, palm uppermost, on the floor: the other played with the tassel of his spear, which stood ever by his side. Suddenly there was a crash at the door, and the sleeper opened his eyes dreamily.

"It is only the old ram, brother: he wants to come in," said Illugi, "and is butting at the door."

"He butts hard; he butts hard!" muttered Grettir: and at that moment the door burst open. They saw faces looking in.

Illugi sprang to his feet, grasped a sword, flew to the doorway, and defended it valiantly, so that none could come within a spear's-length of it; for the lad brought down his weapon on their lances, and smote off the heads.

Then some of the men clambered up on the roof, and began to rip off its covering of turf. Grettir tried to rise to his feet, but could only stagger to his knees. He seized his spear, and drave it through the roof among those who were tearing it down. It struck Karr in the breast, and pierced him.

"Be careful," cried the Hook, — "be careful, and no harm can happen to you!"

Then the men pulled at the gable-ends; heaved the ridge-piece aside, and broke it asunder, so that a shower of rafters and turfs fell into the chamber.

Grettir drew his sword, and smote at the men as they

leaped upon him from the wall. With one blow he struck Vikarr, the servant of Hjalti, over the left shoulder, as he was upon the point of springing down. The sword sliced through him, and came out below his right arm; and the corpse dropped upon Grettir. The blow was so violent, that Grettir fell forward; and, before he could raise himself, Thorbjorn Hook struck him between the shoulders, and made a fearful wound.

Then cried Grettir, "Bare is man's back without brother behind it!" and instantly Illugi threw his shield over him, planted a foot either side of him as he lay on the floor, and defended him gallantly; so that all were amazed at his courage.

"Who showed you the way to the island?" asked Grettir of the Hook.

"Christ showed us the way," answered Thorbjorn.

"Nay, nay!" muttered Grettir. "It was that hag, your foster-mother, who directed you hither!"

The mist of death was in his eyes. He attempted to raise himself, but sank again on the sheepskins, which were now drenched in blood. No one could touch him; for the brave lad warded off every blow that was aimed at his brother. Then the Hook ordered his men to form a ring around them, and to close in on them with shields and beams. They did so; and Illugi was taken and bound, but not till he had wounded the majority of his opponents, and killed three of Thorbjorn's churls.

"You are a brave fellow!" said the Hook; "and never have I seen one of your age who fought so well."

Then they went up to Grettir, who lay in a state of

unconsciousness, without being able to make any resistance.

They dealt him many a blow; but little blood flowed
from the wounds. When all thought that he was dead,
Thorbjorn tried to disengage the sword from his cold,
damp fingers, saying that Grettir had wielded it long
enough.

But the strong man's hand was clinched around the
handle so firmly, that his enemy could not free the
sword from his grasp.

Several of the men came up, and endeavored to un-
weave the fingers; but they were unable to do so. Then
the Hook exclaimed, "Why should we spare this vile
outlaw? Off with his hand!" And they held it
down whilst he hewed it from the arm at the wrist.
Then the muscles of the fingers relaxed; and the
Hook was able to loosen them, and possess himself of
the sword. Standing beside the body, and grasping the
hilt with both hands, he smote at Grettir's head. The
edge of the blade was notched with the blow. "See!"
laughed Thorbjorn: "this mark will be famous in the
history of my sword. I shall show the notch, and say,
'This was done by Grettir's skull!'" He smote twice
and thrice at the outlaw's neck, till the head came off
in his hands.

"Here have I slain a famous warrior!" exclaimed
Thorbjorn. "This head shall come with me to land, that
I may claim the price that has been set upon it, and
that none may be able to deny that I slew the redoubted
Grettir."

The rest of the party told him to do as he chose: but

they did not think much of his act; for they believed Grettir to have been dead before Thorbjorn smote at his head, and they suspected that he had wrought his foe's sickness and death by unhallowed means.

Then the Hook turned to Illugi, saying, "It would be a pity that a brave lad like you should die because you have associated yourself with outlaws and evil-doers."

Illugi answered, "At Al-thing you shall be summoned to give an account of this cursed deed, and answer to the charge of witchcraft, which I shall bring against you if I live."

"Listen to me, boy," said the Hook: "lay your hand to my hand, and take a vow never to revenge that which has taken place to-night, and I will give you life and liberty."

"And listen you to me, Thorbjorn," replied Iiugi. "If I survive, but one thought shall occupy my heart night and day; and that will be, how I can best avenge my brother. Now that you know what to expect from me, choose whether I shall live or die."

Thorbjorn took his companions aside to ask their advice; but they shrugged their shoulders, and replied, that, as he had planned the expedition, he must carry it through as he thought best.

"Well," exclaimed the Hook, "I have no fancy for having the young viper ready to sting me wherever I tread. So he shall die!"

Now, when Illugi knew that they had determined on slaying him, he smiled, and said, "You have chosen that course which is most to my mind."

As the day began to dawn, they led him to the east side of the island, and slew him there. It is said that they neither bound his hands nor eyes; and that he looked fearlessly at them as they smote him, and neither winked nor changed color. Then they buried the brothers beneath a cairn; but they took the head of Grettir, and bore it with them to land.

As they rowed home, the thrall Glaum made such outcries, that they were tired of his noise; and, on reaching the mainland, they slew him.

One morning, Thorbjorn Hook rode with twenty men to Bjarg, in the middle frith, with Grettir's head hanging at his saddle-bow. On reaching the house he dismounted, and stalked into the hall, where Grettir's mother was seated with her servant. Thorbjorn flung her son's head at her feet, and sang, —

> " Flitted I from the island,
>   With me the head of Grettir, —
>   That yellow head which women
>   Weep : with it I am standing.
>   Look you ! the peace-destroyer's
>   Head lieth on the pavement ;
>   Look you ! it cannot moulder
>   Now that it well is salted."

The lady sat proudly in her seat, and did not shed a tear; but, lifting her voice in reply, she sang, —

> " Milksop ! no less than sheep
>   Flee before the fox :
>   Would you have fled before
>   Grettir strong and hale ? "

After this the Hook returned home; and folk wondered at Asdisa, saying that none but she could have borne such sons as those twain who slept in Drángey.

The next day (the 6th) we went back to Akureyri, and slept that night on board "The Curlew."

On the 7th we gave a "final feast" in honor of young Havisteen, for whom we had contracted a warm friendship. It was really hard to bid him good-by. He made us promise to visit him at Copenhagen the next spring; for we had already confided to him — what I now confide to the reader — our intention of visiting Europe during the approaching winter.

On the 8th we sailed for home, and on the 13th had our last glimpse of the Snæfel, — the same white knob low down on the horizon that we had first seen on the 9th of June.

Farewell to the land of ice and fire!

During the homeward voyage, I have been busied in writing out this humble account from the notes I made during our tour. "Would it were worthier!" as somebody says; but I can't help it now, reader. You must take it as it is, if you take it at all; which I respectfully leave with you to elect.

P. S. — Wash begs me to append the following Icelandic song, a great favorite of his. He has been singing it to us all the voyage. I hope his translation of it is not a *plagiarism;* though I strongly suspect it. Raed says he knows it is, and can show me where he stole it. I refused to look, and therefore cannot state.

SWEETLY SWANS ARE SINGING.

Sweetly swans are singing
In the summer-time:
Let us lightly laugh and play,
   Lily maiden!
Sweetly swans are singing.

"Rede my dream right, mother mine,
In the summer-time:
I will give thee golden shrine,
   Lily maiden!
Sweetly swans are singing.

"First methought the moon did smile,
In the summer-time,
Softly over Skaney Isle,
   Lily maiden!
Sweetly swans are singing.

"Then methought a rowan-tree
In the summer-time
Louted lowly unto me,
   Lily maiden!
Sweetly swans are singing.

"Then a swan as silver-white,
In the summer-time,
Lay upon my bosom light,
   Lily maiden!
Sweetly swans are singing.

"And I planets twain did see,
In the summer-time,
Lie a-rocking on my knee,
   Lily maiden!
Sweetly swans are singing.

"Next I saw the tide rise fleet
In the summer-time,
Sweeping o'er my little feet,
  Lily maiden !
Sweetly swans are singing."

"As thou saw'st the moon arise
In the summer-time,
Royal husband be thy prize,
  Lily maiden !
Sweetly swans are singing.

"As the rowan bent, I trow,
In the summer-time,
Many folk to thee shall bow,
  Lily maiden !
Sweetly swans are singing.

"As thou claspedst cygnet fair
In the summer-time,
Thou a princely son shalt bear,
  Lily maiden !
Sweetly swans are singing.

"As thou saw'st two planets shine
In the summer-time,
Lovely daughters shall be thine,
  Lily maiden !
Sweetly swans are singing.

"As around thee stole the flood
In the summer-time,
Shall thy lot be ever good,
  Lily maiden !
Sweetly swans are singing.

" This thy dreaming, daughter mine,
In the summer-time :
Keep thyself, thy golden shrine,
     Lily maiden !
Sweetly swans are singing."

FAMOUS STANDARD
JUVENILE LIBRARIES.

ANY VOLUME SOLD SEPARATELY AT $1.00 PER VOLUME

(Except the Sportsman's Club Series, Frank Nelson Series and
Jack Hazard Series.).

Each Volume Illustrated. 12mo. Cloth.

HORATIO ALGER, JR.

THE enormous sales of the books of Horatio Alger, Jr.,
show the greatness of his popularity among the boys, and
prove that he is one of their most favored writers. I am told
that more than half a million copies altogether have been
sold, and that all the large circulating libraries in the country
have several complete sets, of which only two or three volumes are ever on the shelves at one time. If this is true,
what thousands and thousands of boys have read and are
reading Mr. Alger's books! His peculiar style of stories,
often imitated but never equaled, have taken a hold upon the
young people, and, despite their similarity, are eagerly read
as soon as they appear.

Mr. Alger became famous with the publication of that
undying book, "Ragged Dick, or Street Life in New York."
It was his first book for young people, and its success was so
great that he immediately devoted himself to that kind of
writing. It was a new and fertile field for a writer then, and
Mr. Alger's treatment of it at once caught the fancy of the
boys. "Ragged Dick" first appeared in 1868, and ever since
then it has been selling steadily, until now it is estimated
that about 200,000 copies of the series have been sold.

—*Pleasant Hours for Boys and Girls.*

A writer for boys should have an abundant sympathy with them. He should be able to enter into their plans, hopes, and aspirations. He should learn to look upon life as they do. Boys object to be written down to. A boy's heart opens to the man or writer who understands him.

—From *Writing Stories for Boys*, by Horatio Alger, Jr.

RAGGED DICK SERIES.

6 vols.   BY HORATIO ALGER, JR. $6.00

Ragged Dick.   Rough and Ready.
Fame and Fortune.   Ben the Luggage Boy.
Mark the Match Boy.   Rufus and Rose.

TATTERED TOM SERIES—First Series.

4 vols.   BY HORATIO ALGER, JR. $4.00

Tattered Tom.   Phil the Fiddler.
Paul the Peddler.   Slow and Sure.

TATTERED TOM SERIES—Second Series.

4 vols. $4.00

Julius.   Sam's Chance.
The Young Outlaw.   The Telegraph Boy.

CAMPAIGN SERIES.

3 vols.   BY HORATIO ALGER, JR. $3.00

Frank's Campaign.   Charlie Codman's Cruise.
                Paul Prescott's Charge.

LUCK AND PLUCK SERIES—First Series.

4 vols.   BY HORATIO ALGER, JR. $4.00

Luck and Pluck.   Strong and Steady.
Sink or Swim.   Strive and Succeed.

LUCK AND PLUCK SERIES—Second Series.
4 vols. $4.00
Try and Trust.   Risen from the Ranks.
Bound to Rise.   Herbert Carter's, Legacy.

BRAVE AND BOLD SERIES.
4 vols.   BY HORATIO ALGER, JR. $4.00
Brave and Bold.   Shifting for Himself.
Jack's Ward.   Wait and Hope.

NEW WORLD SERIES.
3 vols.   BY HORATIO ALGER, JR. $3.00
Digging for Gold.   Facing the World.   In a New World

VICTORY SERIES.
3 vols.   BY HORATIO ALGER, JR. $3.00
Only an Irish Boy.   Adrift in the City.
        Victor Vane, or the Young Secretary.

FRANK AND FEARLESS SERIES.
3 vols.   BY HORATIO ALGER, JR. $3.00
Frank Hunter's Peril.   Frank and Fearless.
            The Young Salesman.

GOOD FORTUNE LIBRARY.
3 vols.   BY HORATIO ALGER, JR. $3.00
Walter Sherwood's Probation.   A Boy's Fortune.
            The Young Bank Messenger.

RUPERT'S AMBITION.
1 vol.   BY HORATIO ALGER, JR. $1.00

JED, THE POOR-HOUSE BOY.
1 vol.   BY HORATIO ALGER, JR. $1.00

HARRY CASTLEMON.

HOW I CAME TO WRITE MY FIRST BOOK.

WHEN I was sixteen years old I belonged to a composition class. It was our custom to go on the recitation seat every day with clean slates, and we were allowed ten minutes to write seventy words on any subject the teacher thought suited to our capacity. One day he gave out "What a Man Would See if He Went to Greenland." My heart was in the matter, and before the ten minutes were up I had one side of my slate filled. The teacher listened to the reading of our compositions, and when they were all over he simply said : "Some of you will make your living by writing one of these days." That gave me something to ponder upon. I did not say so out loud, but I knew that my composition was as good as the best of them. By the way, there was another thing that came in my way just then. I was reading at that time one of Mayne Reid's works which I had drawn from the library, and I pondered upon it as much as I did upon what the teacher said to me. In introducing Swartboy to his readers he made use of this expression : "No visible change was observable in Swartboy's countenance." Now, it occurred to me that if a man of his education could make such a blunder as that and still write a book, I ought to be able to do it, too. I went home that very day and began a story, "The Old Guide's Narrative," which was sent to the *New York Weekly*, and came back, respectfully declined. It was written on both sides of the sheets but I didn't know that this was against the rules. Nothing abashed, I began another, and receiving some instruction, from a friend of mine who was a clerk in a book store, I wrote it on only one side of the paper. But mind you, he didn't know what I was doing. Nobody knew it ; but one day, after a hard Saturday's work—the other boys had been out skating on the brick-pond—I shyly broached the subject to my mother. I felt the need of some sympathy. She listened in amazement, and then said : "Why, do you think you could write a book like that?" That settled the matter, and from that day no one knew what I was up to until I sent the first four volumes of Gunboat Series to my father. Was it work? Well, yes ; it was hard work, but each week I had the satisfaction of seeing the manuscript grow until the "Young Naturalist" was all complete.

—*Harry Castlemon in the Writer.*

GUNBOAT SERIES.

6 vols.   By Harry Castlemon. $6.00

Frank the Young Naturalist.   Frank before Vicksburg.
Frank on a Gunboat.   Frank on the Lower Mississippi.
Frank in the Woods.   Frank on the Prairie.

ROCKY MOUNTAIN SERIES.

3 vols.   By Harry Castlemon. $3.00

Frank Among the Rancheros.   Frank in the Mountains.
Frank at Don Carlos' Rancho.

SPORTSMAN'S CLUB SERIES.

3 vols.   By Harry Castlemon. $3.75

The Sportsman's Club in the Saddle.   The Sportsman's Club
The Sportsman's Club Afloat.   Among the Trappers.

FRANK NELSON SERIES.

3 vols.   By Harry Castlemon. $3.75

Snowed up.   Frank in the Forecastle.   The Boy Traders.

BOY TRAPPER SERIES.

3 vols.   By Harry Castlemon. $3.00

The Buried Treasure.   The Boy Trapper.   The Mail Carrier.

ROUGHING IT SERIES.

3 vols.   BY HARRY CASTLEMON. $3.00

George in Camp.   George at the Fort.
George at the Wheel.

ROD AND GUN SERIES.

3 vols.   BY HARRY CASTLEMON. $3.00

Don Gordon's Shooting Box.   The Young Wild Fowlers.
Rod and Gun Club.

GO-AHEAD SERIES.

3 vols.   BY HARRY CASTLEMON. $3.00

Tom Newcombe.   Go-Ahead.   No Moss.

WAR SERIES.

6 vols.   BY HARRY CASTLEMON. $6.00

True to His Colors.   Marcy the Blockade-Runner.
Rodney the Partisan.   Marcy the Refugee.
Rodney the Overseer.   Sailor Jack the Trader.

HOUSEBOAT SERIES.

3 vols.   BY HARRY CASTLEMON. $3.00

The Houseboat Boys.   The Mystery of Lost River Cañon.
The Young Game Warden.

AFLOAT AND ASHORE SERIES.

3 vols.   BY HARRY CASTLEMON. $3.00

Rebellion in Dixie.   A Sailor in Spite of Himself.
The Ten-Ton Cutter.

THE PONY EXPRESS SERIES.

3 vol.   BY HARRY CASTLEMON. $3.00

The Pony Express Rider.   The White Beaver.
Carl, The Trailer.

EDWARD S. ELLIS.

EDWARD S. ELLIS, the popular writer of boys' books, is a native of Ohio, where he was born somewhat more than a half-century ago. His father was a famous hunter and rifle shot, and it was doubtless his exploits and those of his associates, with their tales of adventure which gave the son his taste for the breezy backwoods and for depicting the stirring life of the early settlers on the frontier.

Mr. Ellis began writing at an early age and his work was acceptable from the first. His parents removed to New Jersey while he was a boy and he was graduated from the State Normal School and became a member of the faculty while still in his teens. He was afterward principal of the Trenton High School, a trustee and then superintendent of schools. By that time his services as a writer had become so pronounced that he gave his entire attention to literature. He was an exceptionally successful teacher and wrote a number of text-books for schools, all of which met with high favor. For these and his historical productions, Princeton College conferred upon him the degree of Master of Arts.

The high moral character, the clean, manly tendencies and the admirable literary style of Mr. Ellis' stories have made him as popular on the other side of the Atlantic as in this country. A leading paper remarked some time since, that no mother need hesitate to place in the hands of her boy any book written by Mr. Ellis. They are found in the leading Sunday-school libraries, where, as may well be believed, they are in wide demand and do much good by their sound, wholesome lessons which render them as acceptable to parents as to their children. All of his books published by Henry T. Coates & Co. are re-issued in London, and many have been translated into other languages. Mr. Ellis is a writer of varied accomplishments, and, in addition to his stories, is the author of historical works, of a number of pieces of popular music and has made several valuable inventions. Mr. Ellis is in the prime of his mental and physical powers, and great as have been the merits of his past achievements, there is reason to look for more brilliant productions from his pen in the near future.

DEERFOOT SERIES.

3 vols.   BY EDWARD S. ELLIS. $3.00

Hunters of the Ozark.   The Last War Trail.
Camp in the Mountains.

LOG CABIN SERIES.

3 vols.   BY EDWARD S. ELLIS. $3.00

Lost Trail.   Footprints in the Forest.
Camp-Fire and Wigwam.

BOY PIONEER SERIES.

3 vols.   BY EDWARD S. ELLIS. $3.00

Ned in the Block-House.   Ned on the River.
Ned in the Woods.

THE NORTHWEST SERIES.

3 vols.   BY EDWARD S. ELLIS. $3.00

Two Boys in Wyoming.   Cowmen and Rustlers.
A Strange Craft and its Wonderful Voyage.

BOONE AND KENTON SERIES.

3 vols.   BY EDWARD S. ELLIS. $3.00

Shod with Silence.   In the Days of the Pioneers.
Phantom of the River.

IRON HEART, WAR CHIEF OF THE IROQUOIS.

1 vol.   BY EDWARD S. ELLIS. $1.00

THE NEW DEERFOOT SERIES.

3 vols.   BY EDWARD S. ELLIS. $3.00

Deerfoot in the Forest.   Deerfoot on the Prairie.
Deerfoot in the Mountains.

J. T. TROWBRIDGE.

NEITHER as a writer does he stand apart from the great currents of life and select some exceptional phase or odd combination of circumstances. He stands on the common level and appeals to the universal heart, and all that he suggests or achieves is on the plane and in the line of march of the great body of humanity.

The Jack Hazard series of stories, published in the late *Our Young Folks*, and continued in the first volume of *St. Nicholas*, under the title of "Fast Friends," is no doubt destined to hold a high place in this class of literature. The delight of the boys in them (and of their seniors, too) is well founded. They go to the right spot every time. Trowbridge knows the heart of a boy like a book, and the heart of a man, too, and he has laid them both open in these books in a most successful manner. Apart from the qualities that render the series so attractive to all young readers, they have great value on account of their portraitures of American country life and character. The drawing is wonderfully accurate, and as spirited as it is true. The constable, Sellick, is an original character, and as minor figures where will we find anything better than Miss Wansey, and Mr. P. Pipkin, Esq. The picture of Mr. Dink's school, too, is capital, and where else in fiction is there a better nick-name than that the boys gave to poor little Stephen Treadwell, "Step Hen," as he himself pronounced his name in an unfortunate moment when he saw it in print for the first time in his lesson in school.

On the whole, these books are very satisfactory, and afford the critical reader the rare pleasure of the works that are just adequate, that easily fulfill themselves and accomplish all they set out to do.—*Scribner's Monthly.*

JACK HAZARD SERIES.

6 vols.   BY J. T. TROWBRIDGE.   **$7.25**

Jack Hazard and His Fortunes.   Doing His Best.
The Young Surveyor.   A Chance for Himself.
Fast Friends.   Lawrence's Adventures.

——

International Bibles

Are known the world over for their clear print, scholarly Helps and absolutely flexible bindings. They comprise every variety of readable type in every style of binding and include Text Bibles, Reference Bibles, Teachers' Bibles, Testaments, Psalms, Illustrated Bibles; also the "International" Red Letter Testaments and Red Letter Bibles with the prophetic types and prophecies relating to Christ in the Old Testament printed in red, and the words of Christ in the New Testament printed in red; also Christian Workers' Testament and Christian Workers' Bible in which all subjects or the Theme of Salvation are indexed and marked in red.

For sale by all booksellers. Catalog of Books and Bibles mailed on application to the publishers.

————

THE JOHN C. WINSTON CO.

Winston Building

PHILADELPHIA, PA.

www.ingramcontent.com/pod-product-compliance
Lightning Source LLC
Chambersburg PA
CBHW021517210326
41599CB00012B/1288

*9 7 8 3 3 3 7 3 1 7 9 6 6 *